D0320938

CLINICAL TEACHING: A GUIDE TO TEACHING PRACTICAL ANAESTHESIA

CLINICAL TEACHING: A GUIDE TO TEACHING PRACTICAL ANAESTHESIA

EDITORS

DR J DAVID GREAVES
Royal Victoria Infirmary, Newcastle upon Tyne, UK

PROFESSOR CHRIS DODDS
James Cook University Hospital, Middlesbrough, UK

DR CHANDRA M KUMAR
James Cook University Hospital, Middlesbrough, UK

PROFESSOR BEREND METS
Milton S Hershey Medical Center, PA, USA

LONDON AND NEW YORK

Library of Congress Cataloging-in-Publication Data

(*Applied for*)

Cover design: Studio Jan de Boer, Amsterdam, The Netherlands
Typesetting: Charon Tec Pvt. Ltd, Chennai, India

Published by: Taylor & Francis
2 Park Square, Milton Park, Abingdon, Oxon, OX14 4RN
270 Madison Ave, New York NY 10016
www.taylorandfrancis.com

Transferred to Digital Printing 2006

ISBN 90 265 1941 9

Publisher's Note
The publisher has gone to great lengths to ensure the quality of this reprint but points out that some imperfections in the original may be apparent

Printed and bound by CPI Antony Rowe, Eastbourne

Contents

Editors and Contributors VII
Foreword IX
Introduction XIII

Section 1 Setting the scene for learning **1**

1. Providing the right environment for learning 3
David Greaves
2. Clinical supervision 13
David Greaves
3. Learning from work 21
David Greaves
4. The assessment of clinical competence 33
David Greaves
5. The ethics of learning on patients 45
Catherine Bartley
6. The non-technical skills of anaesthetists 53
Georgina Fletcher and Ronnie Glavin
7. Problem-based learning (PBL) 63
Philip Liu and Letty M. P. Liu
8. Educational supervision and mentoring 71
Chandra Kumar and Chris Dodds
9. Learning by maintaining a 'Portfolio' 83
Shashi Kant Gupta and David Greaves

Section 2 Clinical teaching **91**

10. Teaching anaesthesia in the operating theatre 93
David Greaves
11. Helping trainees to develop decision-making skills 103
David Greaves and Michael Olympio
12. Teaching practical procedures 121
David Greaves

13. Informal mini-tutorials in the operating theatre 133
David Greaves

14. Teaching anaesthesia to new starters 143
David Greaves

15. Teaching anaesthetists how to behave in an acceptable professional manner 155
David Greaves

16. Making routine judgements about clinical competence 165
David Greaves

17. Giving feedback and monitoring progress 183
Berend Mets

Section 3 Using simulators for teaching **195**

18. An introduction to simulation in anaesthesia 197
Ronnie Glavin and Nicki Maran

19. Simulation and technical skills 207
Ronnie Glavin and Nicki Maran

20. Simulation and non-technical skills 219
Ronnie Glavin and Nicki Maran

21. Setting up a high fidelity simulator centre 231
Ronnie Glavin and Nicki Maran

22. Practicalities of simulation: adding role-play to low and intermediate fidelity simulation 241
Aidan Byrne and Matthew Checketts

23. How to organise a major obstetric haemorrhage 'fire-drill' 253
Rachel Walpole and Vicki Clark

Appendix 1 A guide for teaching moderators how to conduct a problem based learning session 259

Appendix 2 Anaesthesia for empyema drainage in a patient with a broncho-pleural fistula 265

Index 271

Editors and Contributors

Editors

DR DAVID GREAVES	Consultant Anaesthetist	Royal Victoria Infirmary, Newcastle Upon Tyne, UK
PROFESSOR CHRIS DODDS	Consultant Anaesthetist	James Cook University Hospital, Middlesbrough, UK
DR CHANDRA KUMAR	Consultant Anaesthetist	James Cook University Hospital, Middlesbrough, UK
PROFESSOR BEREND METS	Eric A Walker Professor and Chair	Dept of Anesthesiology, Milton S. Hershey Medical Center, Hershey, PA, USA

Contributors

DR CATHERINE BARTLEY	Consultant Anaesthetist	Queen Elizabeth Hospital, Gateshead, UK
DR AIDAN BYRNE	Consultant Anaesthetist	Morriston Hospital, Swansea, UK
DR MATHEW CHECKETTS	Consultant Anaesthetist	Ninewells Hospital, Dundee, UK
DR VICKI CLARK	Consultant Anaesthetist	Simpson Centre for Reproductive Health, Edinburgh, UK
MS GEORGINA FLETCHER	Research Fellow	Industrial Psychology Group, Dept of Psychology, University of Aberdeen, King's College, UK
DR RONNIE GLAVIN	Consultant Anaesthetist	Scottish Simulator Centre, Stirling, UK
SHASHI KANT GUPTA	Research Fellow	Dept of Education, University of Newcastle, UK
DR LETTY M.P. LIU	Professor	University of Medicine and Dentistry of New Jersey – New Jersey Medical School, Newark, NJ, USA

DR PHILIP LIU	Chair of Anaesthesiology	University of Medicine and Dentistry of New Jersey – New Jersey Medical School, Newark, NJ, USA
DR VICKI MARAN	Consultant Anaesthetist	Scottish Simulator Centre, Stirling, UK
DR MICHAEL OLYMPIO	Professor of Anaesthesiology	Director, Patient Simulation Laboratory, Wake Forest University School of Medicine, Winston-Salem, NC, USA
DR RACHEL WALPOLE	Consultant Anaesthetist	Royal Gwent Hospital, Newport, UK

Foreword

Dr Douglas Justins

Department of Anaesthetics, St Thomas' Hospital, London

The old version of the Hippocratic Oath included an undertaking to impart knowledge of the art of medicine. A recently revised version says that doctors should be ready to share their knowledge by teaching others what they know. In the United Kingdom the General Medical Council encourages doctors to contribute to the education and training of other doctors, medical students and non-medical healthcare professionals. This obligation can be extended to the need to supply information and education for the public.

Teaching in some form or another is specified in virtually every job description for consultants in the National Health Service but, so far, there has not been any move to stipulate possession of additional formal qualifications in medical education. It is not inconceivable that in the future the medical Royal Colleges, or the Postgraduate Deans, will have to create a system that gives formal certification to doctors who wish to have responsibility for teaching. The hospitals and organisations that pay for medical education will wish to see that the money they spend is being used to maximum effect by people who know what they are doing.

A medical qualification and ability to practice good medicine does not guarantee an ability to teach the elements of good medical practice to others. Some doctors assume that the ability to teach is conferred along with the basic medical degree but this attitude is fast changing. This is not to say that everything that doctors have been doing has been wrong and most still do a very good job of teaching but with additional training these skills can be improved. Now it is recommended that doctors who accept special responsibilities for teaching should develop and maintain the skills, attitudes and practices of a competent teacher.

Effective teaching is not just a matter of giving formal lectures but instead involves many other facets. Planning of teaching requires careful organisation and the creation of an environment that is conducive to learning. The trainees require defined learning objectives. The teacher should be familiar with a range of techniques including small group teaching, formal lecturing and one-to-one teaching. Teaching in the operating theatre is a very important component of training in anaesthesia. Many of the skills and attitudes of a professional life will be forged during this one-to-one teaching in theatre. The teacher will be acting as a role model for the trainee. In addition this is the environment in which the trainee learns many of the practical techniques that are very important in anaesthesia, critical care and pain management. The effective teacher will be responsible for preparing teaching material. This may range from simple tutorial notes right through to innovative methods of electronic learning.

The preparation of candidates for examinations provides a focus and a stimulus for trainers and trainees but it is now recognised that a commitment to life-long learning is a corner stone of professionalism. The good teacher will provide motivation and maintain trainee involvement in learning. The good teacher will instil into the trainees the desire to use education and learning to underpin the safe practice of medicine throughout their careers. The doctor will have to learn to sieve out the important and relevant bits of evidence based medicine. They will have to extract maximum value from continuing education events. Doctors who wish to keep up to date or acquire additional qualifications may participate in distance learning or web based learning.

The supervision of training has assumed an increasingly important place in medical education and in the broader context of risk management for organisations that provide training. Less experienced doctors and students must be properly supervised so that the welfare of patients is never compromised. Public, political and legal opinion has changed considerably of late and patient safety is now a central tenet of the health service.

Trainers are responsible for monitoring progress and guiding the direction of training through a system of assessment and appraisal. It is essential for trainers to understand the theory and practice of assessment and appraisal. They must be able to use assessment and appraisal for the benefit of the trainee. They must become accomplished at giving feedback. Some become involved in mentoring schemes. In addition it is highly beneficial for trainees to understand the processes through which they must progress because regular assessment is an integral part of the competence based training programme for anaesthesia in the UK. Ultimately the educational supervisors must take responsibility for certifying the competence of individual trainees thereby allowing them to progress to independent practice. Established career grade doctors now face revalidation to ensure that they maintain a satisfactory standard of practice. To be fair and reliable this process must be conducted by individuals who have a sound understanding of educational principles. Certification or revalidation of doctors who are not competent will put patients at risk in anaesthesia, critical care and pain management.

The assessors must get it right for the sake of the individual and for the sake of the health-service.

Medical education is facing new challenges. The last few years have been witness to major reforms in postgraduate medical education in the UK. Anaesthesia has introduced a new competency based training programme with well-defined assessment gates along the way. In many cases the time available for training and education has been reduced by the reforms and by the reduction in junior doctors' hours. The time that is available will have to be used more efficiently and effectively. There is no place for a haphazard approach to the organisation or delivery of medical education. Trainees now undertake a reduced workload. The old apprenticeship scheme of learning by high-volume experience has disappeared. Supervision is more stringent and the premature assumption of clinical responsibility belongs to the pre-litigious era of the health service. The reduced time available for training may be further eroded by new working time directives and by increasing political pressure to ensure that service takes precedence over training. Trainers and trainees will have to devise ways of extracting the maximum training benefit from the service component of the trainee's job. Simulators will replace patients. The increasing complexity of medicine will make new demands on doctors. Already there is increasing specialisation in anaesthesia and surgery. Also, doctors play a major role in health service management in addition to managing clinical care and medical education.

Teachers and educational supervisors will have to possess specific professional and personal qualities if they are to meet these new challenges. The performance of teachers and educational supervisors will have to be monitored so that problems or deficiencies can be addressed. Initial training in medical education will have to be linked to on-going programmes that will enable doctors to remain competent to undertake a range of different teaching responsibilities.

The knowledge, skills and attitudes of a good teacher can be acquired from a number of sources. Books such as this volume and journals provide a fruitful source of knowledge. There are some excellent courses run by those responsible for the delivery of education. In the UK this includes postgraduate deans, universities and medical Royal Colleges. Distance learning courses in medical education are available from a number of universities. The Royal College of Surgeons has been running 'Training the Trainers Courses' since 1994. The Royal College of Physicians has launched a programme called 'Physicians as Educators'. Accreditation as an 'RCP Educator' is offered by the scheme. The Royal College of Anaesthetists runs a series of 'How to Teach' courses and in association with the Centre for Medical Education, University of Dundee has introduced a Certificate in Medical Education for Anaesthetists. There are thriving national and international organisations devoted to medical education. The contributors to this volume have wide experience in medical education and many have made major contributions to teacher training initiatives in the UK and in other countries.

This volume is aimed at trainees as well as career grade doctors and experienced teachers. Training in medical education should not be restricted to more senior doctors. This knowledge will help trainees to learn how to learn as well as how to teach. Senior trainees teach junior trainees and are present during some of the most important and formative episodes in a starter trainee's development. Often these episodes occur during on-call duties. The new training programme for anaesthetists acknowledges the importance of medical education by including it as a specific generic topic for the latter years of training. Medical education accompanies topics such as information technology, healthcare management, research methodology and medical ethics.

As a doctor you should be looking to enhance your personal portfolio, expand your skills and to explore new fields. Medical education is very rewarding and you can use participation in teaching to underpin your own continuing development. Exploring new teaching skills can provide a stimulating intellectual challenge. Improvement of your practical teaching skills and a greater understanding of the principles of education as applied to medicine will assist you to develop both as a doctor and as a teacher. Effective teaching will have an enduring impact on patient care long after you cease direct patient contact.

Douglas Justins
26 March 2002

Introduction

There is no textbook available that can lead and inform the established trainer or trainee on the practical and educational aspects of learning the practice of anaesthesia. Yet this is one of the greatest areas of concern to new specialists and senior trainees. Even experienced doctors are worried by their responsibility for teaching and often feel that they are poor teachers. Usually these fears are groundless. The bedrock of learning in medicine is clinical teaching in the workplace and in this arena consultants are the experts. There are however many ways to lend purpose and structure to clinical teaching and make it relevant to the individual learner. In the classroom and lecture hall a little thought and preparation can make the session more interesting and maximise the benefit to the learners.

Until ten or fifteen years ago doctors behaved as though good medical teachers were possessed of an innate talent. It is now accepted that few doctors are 'naturals' and that everyone's teaching can be improved by a little thought and preparation. Today there are a number of books, courses and degree programmes that will help doctors to improve their classroom teaching and lectures. Most of the learning in clinical medicine is informal, practical and occurs in the workplace. Much of this is not recognised to be teaching either by the teacher or the learner. Clinical teaching is, however, of great importance. It is axiomatic that no matter what you say the learner is most likely to copy what you do. Despite this, medical teachers consistently model poor behaviour and flawed values and if their acolytes see that these strategies are effective they will be apt to adopt them and perpetuate unacceptable clinical practice into the next generation. The inexperienced doctor is confronted by a tangle of interconnected learning tasks in the wards and operating theatres. Experienced doctors recognise patterns and priorities amongst these but seldom understand the importance of discussing their learner's insights, which may be seriously awry, or of helping them see order in chaos.

It is of vital importance to all of us how doctors learn medical practice. Insufficient is known about how quickly we learn practical medicine, what additional complications result from the work of learners and what effect different ways of learning and supervising can have on these. The existing literature suggests that the ways of learning used in UK hospitals lead to 'learning curve morbidity' being a major contributor to total patient morbidity. Is there anything the individual teacher and learner can do about this? We believe there is.

This book is intended for the trainee as much, if not more, than for teachers. Learners have learning high on their list of personal priorities and our experience with a postgraduate qualification for anaesthetic teachers has shown that the trainees are very interested in learning to teach. Clearly it is worth leading a thirsty horse to the water. Secondly, we need to learn to learn. As doctors we have learned and have been successful, but the principle obstacles to our career progress at school and in medical school have been the ability to understand difficult principles and to memorise vast numbers of facts. Throughout our clinical career our effectiveness to patients will depend substantially upon a different type of learning. This differs significantly in that the learner decides his or her learning needs, determines his or her own curriculum and self-assesses the outcomes of learning. In addition, the workplace calls for communication, negotiation and organisational skills that have not been taught at all. We hope that trainees will use this book to help understand their own learning project for professional life.

Clinical teachers are required to assess trainees so that they can comment upon their progress and to match their teaching to the trainee's progress. This is an area where consultants need guidance. In this text is advice about making judgements about the learner's progress. Knowing what to look for the teacher will be able to make useful comments about the trainee's performance. Clinical assessment has become embroiled in a confusing terminology that must be understood. Often, as in the case of competencies, the educational terms used have simple associations with everyday vocabulary. Therein lies a problem because the meaning of neologisms such as competency cannot be inferred from their association with competence. Competence, appraisal, assessment all have specific technical meanings in the context of learning. It is not surprising to find that many practising clinicians are confused about what their assessments mean and unsure how to arrive at judgements. This is a very serious issue when training schemes rely heavily upon professional, in-service assessment by large numbers of supervising specialists.

This book is intended for all those who teach and all those who learn the trade of anaesthesia. The clinical practice within specialties is sufficiently diverse to mean that clinical teaching is in its detail specialty specific and above all we have sought to get down to that detail. We are hoping to help practical teachers and learners. Wherever possible specific 'lesson plans' are given to suggest ways to organise the work and learning. A trainee in the clinical workplace has pressing need to learn and the teacher has a willing audience. Emphasis is laid upon relevance; the relevance of the teaching to the clinical situation and the relevance to the needs of the learner. It has been remarked that the human species is an extraordinarily adaptive learner when need arises and postgraduate medical learners have pressing needs. Good teachers seize this opportunity.

The book is organised into three sections. In the first section we concentrate principally on the necessary physical and intellectual resources needed for good clinical teaching. In the second section we discuss specific aspects of clinical teaching and in the third section we introduce the concepts of teaching anaesthesia using proxies for the clinical problem — simulation and role-play. In all sections we have provided only as much discussion of theory as we believe is necessary for insightful practice. It must be stressed that our colleagues, our friends, our associates and ourselves have developed the principles of practical teaching that we espouse during many, many hours of experience in the operating theatre.

The Editors

Section 1

Setting the scene for learning

1

Providing the right environment for learning

David Greaves

In 1988 Steven Jonas wrote in the forward to one of the first textbooks devoted to clinical teaching, "There are two principal influences on physician behaviour: how they are paid and how they are trained."[1] Looking round the world we find that there are striking differences in the way anaesthetists are trained; not least in the duration of training, which ranges from two to seven years. Some countries have in-service exams, some have an exit exam and others have no exams (see Table 1).[2] We must conclude that there

"And so I finally realised that, after all, my true calling was to be an anaesthetist."

Table 1. Training round the world. (Adapted from Pal and Strunin[2], with permission.)

Country	Years	Research & publications	Entry requirements	Exam	Exit exam
UK & Ireland	7	Not essential	1 year post-graduation; interview	Yes	No
Austria	6	Not essential	Basic Medical Qualification	No	No
Hong Kong	6	Not essential	Basic Medical Qualification	Yes	No
Singapore	6	Essential	2 year post-graduation	Yes	Yes
Australia	5	Essential	1 year post-graduation	Yes	No
Belgium	5	Essential	Basic Medical Qualification; interview or 2 month evaluation	Yes	Yes
Germany	5	Essential	1 year post graduation	Yes	Yes
Netherlands	5	Essential	Basic Medical Qualification	Yes	Yes
New Zealand	5	Essential	1 year post-graduation	Yes	No
Norway	5	Not essential	18 months internship	No	No
Poland	5	Essential	Basic Medical Qualification; by agreement of Head of Department	Yes	Yes
Portugal	5	Not essential	Basic Medical Qualification; entrance test (MCQ)	Yes	Yes
Romania	5	Not essential	Basic Medical Qualification; entrance test	Yes	Yes
Zimbabwe	5	Dissertation	Basic Medical Qualification	Yes	Yes
Sri Lanka	4½	Essential	Full Registration	Yes	Yes
Canada	4	Not essential	Basic Medical Qualification; 1 year internship	Yes	Yes
China	4	Essential	Basic Medical Qualification	Yes	Yes
France	4	Final thesis	Basic Medical Qualification; entrance test (internal)	Yes	Yes
India	4	Essential	1 year post-internship; entrance test (MCQ)	Yes	No
Italy	4	Not essential	Basic Medical Qualification; test	Yes	Yes
Lebanon	4	Essential	Internship	Yes	Yes
Macedonia	4	Diary of anaesthetics	1 year post-internship	Yes	Yes
Nigeria	4	Essential	Full Registration	Yes	Yes
Papua New Guinea	4	Essential	Basic Medical Qualification; candidates' choice	Yes	Yes
South Africa	4	Not essential	1 year post-internship	Yes	Yes
USA	4	Not essential	Basic Medical Qualification; 1 year internship	Yes	Yes
West Indies	4	Thesis or Diary	Internship	Yes	Yes
Iran	3	Essential	Basic Medical Qualification; entrance test	Yes	Yes
Syria	3	Essential	>70% marks in MBBS	Yes	Yes
Ukraine	3	Not essential	Basic Medical Qualification	Yes	Yes
Uruguay	3	Not essential	Pass introductory course	Yes	Yes
Brazil	2	Essential	Basic Medical Qualification	Yes	Yes
Fiji	2	Not essential	2 year post-internship	No	No
Iceland	2	Not essential	Internship	No	No
Belarus	1	Not essential	Internship	No	No

is a lack of consensus about how best to train anaesthetists. The differences run even deeper. There is no consensus as to what a specialist anaesthetist is. In the UK there is one grade of consultant but the job may include sessions in sub-specialties such as critical care and pain management to which some countries devote entire training schemes. Some countries have several grades of specialist with the possibility of promotion with further experience. In the United States, practice and habits of work are divided between university academic and independent institutions. Clearly there are differences in the approach to training that depend on these factors but the educational project is similar wherever anaesthetists train. This text is devoted to those similarities.

The Anaesthetic Department

The Anaesthetic Department should be an open, welcoming place, at the focus of professional life for consultants and trainees alike. It should be a meeting place, a busy hive of activity to which trainees will gravitate in their spare time between lists and

- Make the departmental base a friendly welcoming place.
- Be alert to the effects of personal problems adversely affecting clinical performance.
- Be on the lookout for work-related stress.
- Encourage trainees and consultants to use the department for rest and relaxation.

at the start and end of the day. Learning anaesthesia and passing professional examinations is hard and often lonely work. An anaesthetic department that has a businesslike air of education sets the right tone for learning. Anaesthesia is a hazardous business and not only for the patient. Personal problems can develop from strains in the workplace and problems at home can seriously lower performance in the operating theatre. In a friendly and supportive department someone is likely to know that a trainee is having difficulty. It is then possible to act, helping the trainee and keeping the patients safe, before matters come to a crisis.

A workplace in which trainees and consultants get to know one another has professional as well as social benefits. Trainees must be able to turn to consultants and ask for help knowing that they will not be destructively criticised or ridiculed. Everyone is more confident of a sympathetic hearing from colleagues they have got to know personally. We have all needed to talk through some mishap or unpleasant professional experience and the group can be very useful in helping resolve anger, guilt and feelings of failure. It is instructive to listen to a trainee informally recounting a problem to and it is surprising to find how readily they will accept their colleagues' view that they should have managed matters differently.

The department must have good facilities; a communal sitting area, big enough and comfortable enough for people to use, a room and equipment for eating, cooking and making beverages and a meeting room big enough for tutorials.

When appointing a departmental secretary or manager, consideration should be given to their role. This includes looking after the facilities, ordering and maintaining equipment and providing secretarial services for the trainees. The job advert should refer to this organising role and the appointment committee should take note of the candidate's personal qualities.

A good department will have a social life. An occasional night out, sporting event and an annual Christmas party can play a major role in developing a sense of community. Some trainees and consultants deplore such activities and actively oppose them. They ignore the educational and work advantages of fusing the anaesthetic department into a team.

The trainees role in the department — how to get the best out of the experience

The trainee also has responsibilities as a member of an anaesthetic department. I have dwelt on the importance of the anaesthetic community and the trainee is as much responsible for this as the consultants. A trainee has a responsibility to their peers.

The teaching in a department depends upon everyone making an effort. The trainee should take the trouble to get to know the seniors in their department. I have advocated this to consultants but trainees must be prepared to meet them halfway.

The trainee has a particular responsibility for formal teaching. When consultants organise small-group teaching, journal clubs etc it is essential that trainees attend whenever possible. There is a striking difference between departmental teaching in the UK and the USA. In the USA teaching is organised very early in the morning but is well attended. In the UK trainees will often find excuses not to attend even when the teaching is in protected time. This discourages the consultants and they stop organising teaching. I am not in favour of coercion for mature adult learners but everyone must understand that if they ignore the available teaching the educational quality of their programme will go downhill.

Trainees must also avoid becoming dependent. People have individual learning styles and each trainee must understand what works best for them and be positive in searching it out. At an early stage the trainee should begin to plan how they intend to learn, when they intend to take professional exams, what sort of sub-specialty anaesthetics they incline towards and how they intend to learn about disciplines such as clinical research and audit. At the outset of a new appointment they should read the relevant curriculum, set themselves targets and work out a timetable. Get to know what special experiences are available and organise them for yourself.

Advice for trainees

- Take pride in your work.
- Be active in the life of your department.
- Co-operate with your colleagues about list placements, on call and annual leave.
- Be in control of your own learning.
- Set yourself learning targets.
- Learn about practical teaching and learning.
- Be pro-active – ask the consultants how you are doing. Seek criticism as well as praise.
- Be pro-active – tell the consultants what is good and bad about the placement.
- Be pro-active – ask for the experience you need.

Consultants as teachers

Learning *should* be exciting and doctors who teach must always try to interest, extend and challenge in order to take advantage of the enthusiasm that the learner

Some characteristics of good clinical teachers

- Enthusiasm
- An interest and willingness to learn themselves
- Good knowledge of the material
- Good communication skills
- Competent performers in clinical care
- Intellectual honesty
- Respect for trainees
- Flexible enough to allow trainees to do things their own way
- Good humoured
- Able to engage in effective case-based teaching
- Uses question and answer teaching at higher taxonomic levels of learning (see Chapter 11)

brings with them at the start of their training. Consultants are often unsure of their ability to teach, but, at least with regard to clinical teaching, they need not worry.

One to one theatre-teaching is at the heart of anaesthetic learning and every consultant has a wealth of experience of it. They have learned the skills of practical teaching in the course of their own time as learners. The operating theatre teaching routine consists of consultants demonstrating practice and then watching as the trainee attempts to copy what they have just seen. Trainees take part in thousands of such educational exchanges and become familiar with the routines of demonstration and analysis of performance that lie at the heart of teaching by example. Their experience makes them valuable instructors and they should have confidence in their abilities.

Good teachers are in sympathy with their students. Consultants who are indifferent to trainees will never be charismatic and popular teachers. On the other hand many consultants with a relatively neutral attitude have much to contribute within a

Characteristics of excellent teachers identified by Cleave-Hogg and Benedict[3]

- Dedication
- Enthusiasm for teaching
- Willing to give time to teaching
- Enjoys his or her professional work
- Understands themselves as a role model
- Motivated to continuously update and enrich their own learning
- Able to establish and maintain inter-active professional relationships

Characteristics of attending physicians identified as excellent role models[4]

- Spending additional time with trainees
- Formal training in teaching
- Stresses the importance of doctor patient relationship
- Teaches psychosocial aspects of medicine
- Enjoys teaching trainees
- Learns about the lives of trainees

department's teaching programme. It would be a mistake for them to be discouraged from teaching. Cleave-Hogg has reviewed the characteristics of popular anaesthetic teachers.[3] She found that a group of trainers who had averaged above four on a five-point rating by residents had a number of characteristics in common. They regularly taught using an 'enquiry' approach to learning and they understood the complexity of the anaesthetic task and decision-making. It appeared that trainees appreciate supervisors who challenge them to analyse and evaluate, rather than those who let them work without justifying their choices.

Trainees as learners

Trainees are faced with three types of learning that may at times appear to be in conflict.

- Practical anaesthesia.
- The theoretical basis of everyday practice.
- Preparing for licensure or certification exams.

Anaesthetic trainees are adults and have different characteristics as learners to schoolchildren. Adult learners are usually said to be volunteers and often have strong personal motivation to learn and succeed. Indeed Greenhalgh has suggested that adults

"I hope the children aren't bothering you."

Adult learners

- Have chosen to learn.
- Have the intention to learn.
- Like to have some control of what they learn.
- Like to have choice about how they learn and generally prefer active learning.
- Are goal orientated and like to be able to see the practical use of the things they are learning.
- Learn in a context of a variety of past experiences.
- The past experiences of the learner are a major resource in adult learning situations.
- Adults prefer to be treated as individuals. They are a highly diversified group with widely different preferences, needs, backgrounds and skills.
- Adult learners like to see progress from dependency to independence as they grow in responsibility, experience and confidence.

seek to learn *in order* to bring about personal change.[5] They have attended for the purpose of learning and have the intention to learn. This oversimplifies the situation and it must be understood that some medical postgraduate trainees are making the best of their available choices but are not enthusiastic about having to continue the grind of hard learning.

When it comes to learning practical anaesthesia almost all trainees are strongly motivated, because day-to-day work depends on their mastery of practical techniques and clinical decisions. They find giving their first anaesthetics a terrifying experience and have an urgent need to learn so that they can reduce the stress of work. Nurses, anaesthetic assistants and surgeons watch the new trainee and this can be an acutely uncomfortable experience. He or she really wants to cope and look good. Similar considerations relate to learning the theoretical basis of clinical practice; but motivation to learn for exams is another matter. Traditionally the syllabus for medical examinations contains vast amounts of factual material. An anaesthetic trainee is less keen to learn this sort of material. Every consultant has known competent trainees that have been unprepared to put in the work needed to master their material and have preferred not to complete their training. In practice even an apparently well motivated trainee, who has made short work of the necessary job related learning, may need some help and support to get round to passing the exam.

Novice anaesthetists are strongly motivated to learn practical anaesthetic techniques by their fear of failing to cope and the associated fear of looking stupid. ...

... When they choose anaesthesia, however, they have to accept the whole package and some types of learning are easier than others.

Schoolchildren and university students are full time learners, whereas postgraduate trainees in anaesthesia have large service work commitments. Efficient 'book learning' demands a clear head, so bookwork and long days in theatre are in conflict. The work of trainee anaesthetists must be organised in such a way that they have adequate rest and time for study when they need it.

Adult learners may be confronted by serious personal difficulties that affect both their motivation and the organisation of their exam preparation. In general, trainers cannot solve these problems but they certainly need to know about them. Everyone accepts that emotional difficulty, money worries and so on must not be allowed to affect patient care. Similarly, steps should be taken to minimise their effect on training. Trainees must be helped to explore and resolve problems that are eroding their time for study.

When adult learners come to new areas of knowledge they may revert to learning patterns that are more commonly seen in children. They become less self-directed and prefer material to be presented to them rather than to search for it themselves. This tendency can rapidly lead to a type of 'tutor dependency'. The trainee expects continual goal orientated teaching and is apt to blame the consultants for their own lack of success. Tutor dependency is often seen in trainees who are struggling to pass theory exams. This dependency cycle must be discouraged, as it does not enable the development of self-motivated lifetime-learning habits as the trainee passes to the independence of consultant practice.

References

1. Douglas KC, Hosokawa MC, Lawler FH. *A practical guide to clinical teaching in medicine*. New York: Springer Publishing Co, 1988;xiii.
2. Pal SK, Strunin L. UK structured course of training in anaesthetics: world-wide comparisons. *Current Anaesthesia and Critical Care*. 1997;8:162–166.
3. Cleave-Hogg D, Benedict C. The characteristics of excellent clinical teachers. *Canadian Journal of Anaesthesia*. 1997;44:577–581.
4. Wright SM, Kern DE, Kolodner K, Howard DM, Brancati FL. Attributes of excellent attending physician role models. *New England Journal of Medicine*. 1988;339:1986–1993.
5. Fraser SW, Greenhalgh T. Coping with complexity: educating for capability. *British Medical Journal*. 2001;323:799–803.

2

Clinical supervision

David Greaves

Clinical supervision

Common words tend to take on special meanings in disciplines such as education and supervision has come to mean a variety of specific things. In some fields, for example, it refers to a process of discussion intended to help the worker maintain detachment from the 'client' and their problems. In others it relates to the facilitation of a process of professional development.[1] In anaesthesia we generally understand clinical supervision to be a process by which a specialist ensures that the patient receives proper treatment whilst in the hands of a learner. This guardianship of quality must be conducted in a way that allows the trainee to develop whilst keeping the patient safe. Clinical supervision is distinct from 'educational supervision', the process by which the learner's progress is planned, guided and monitored.

> *In anaesthesia we generally understand clinical supervision to be a process by which a specialist ensures that the patient receives proper treatment whilst in the hands of a learner.*

Supervision keeps patients safe. Consultants must know what their trainee can and cannot do and trainees must be taught when and how to ask for help.

> *Much of what occurs in the operating theatre is predictable and consultants and trainees are able to plan supervision that will keep patients safe.*

Learners are necessarily unskilled and professionally ignorant in some aspects of their work and will therefore make mistakes. These mistakes are immediately harmful to patients. Part of the professionalism of consultants is to ensure that their supervision prevents lapses in the quality of care.

There are major differences in the approach to supervision between countries. In some systems learners seldom, if ever, work without the close presence of a specialist, whilst in others it is accepted for relatively inexperienced anaesthetists to administer anaesthetics with their nearest help outside the hospital. In the United Kingdom very inexperienced trainees sometimes undertake anaesthesia with their closest helper 30 minutes or more away. These trainees administer many anaesthetics without any input whatsoever from a senior colleague and indeed many patients are treated entirely by trainees in all specialties and never see a trained specialist throughout their stay in hospital. Many UK consultants are very unhappy with this situation. Where work practices and the requirements for health insurance funding mandate intimate supervision there is less necessity for the supervising specialists to agonise about their responsibilities.

Supervision must be appropriate

Supervision will be direct and continuous at the outset of anaesthetic training. As the learner improves, the consultant will withdraw supervision and allow the trainee more opportunity to make his own clinical decisions.[2] As trainees move to new areas they are effectively new starters again. In each sub-specialty area the trainee returns to continuous supervision followed by recognition of the elements of competence and a move towards less supervised practice.

How can you tell how much supervision a trainee needs?

Consultant anaesthetists work with many trainees and in some situations they may have very little information about them. Particularly when working on call, out of hours, the patient's safety depends upon the consultant being able to make the right assumptions about the trainee's need for supervision. There are skills and knowledge which novice trainees must have in order to be able to work safely in the on call situation. In the past in the UK, the Royal College of Anaesthetists has recommended that independent practice out of hours is not undertaken in the first six months of anaesthetic training. With the new competency-based training programme the advice is that no specific time limit should be observed but that supervision should be maintained at the closest level until the trainee has been checked for the initial test of competency. Reasonable trainees in good departments should all have achieved the necessary standard by six months. It is prudent to take extra measures to try and ensure that all necessary teaching has been done and that the trainee has achieved a safe level of practice. In the USA levels of supervision are

> *The need for supervision brings consultants and learners together in the operating theatre creating an ideal opportunity for teaching.*

The Royal College of Anaesthetists' definitions of levels of supervision:

- *Immediately available supervision:* the supervisor is actually with the trainee or can be called within seconds.
- *Local supervision:* the supervisor is on the same geographical site, is immediately available for advice and is able to be with the trainee within ten minutes of being called.
- *Distant supervision:* the supervisor is rapidly available for advice but is off the hospital site.

Further expansion of these definitions can be found in the RCA guidance.[2]

usually decided by the specific hospital in conjunction with the residency program in which the trainee working. In general, supervision in the USA is closer than in the

Deciding how closely to supervise a trainee:

1. Ask about the length and intensity of the trainee's general experience in anaesthesia and in the specific work they are now required to undertake.
2. Ask about their examination progress and whether they have done ALS and ATLS courses.
3. Ask the trainee to tell you about the level of supervision that they have previously received for the specific type of work now required. Is the level of independence they claim consistent with their experience? Have they been performing at a level that you would expect of similarly experienced trainees?
4. Ask the trainee what they would do in various situations appropriate to their grade and the work planned.
5. Ask the trainee to tell you about the last occasions that they called for help or advice. Were these requests in line with their stated experience?
6. Ask the trainee to tell you about the sort of circumstances in which they would call you for help.
7. Ask the trainee to suggest some ground rules for their supervision.
8. Decide whether you need to see the trainee work before you can leave them alone.

UK, which is not surprising in light of the fact that the anaesthetic experience component of training lasts eight years, as opposed to three in the USA.

The trainee must keep a logbook and this should be checked against the list of required experience. Doing set numbers of procedures does not ensure skill but mastery cannot be achieved without repetition, so it is reasonable for a department to set minimum numbers for experience prior to being allowed to undertake on call work.

How trainees should approach supervision:

- Have you done this before?
- Do you know what difficulties might arise in this case?
- How likely is it that a problem will arise?
- Have you managed the possible problems previously?
- Are you confident that you will manage this case to a satisfactory standard?
- How far away is help?
- Are you frightened?

Testing progress

Progress can be checked by observing the trainee at work and making an estimate as to whether they are sufficiently competent. This can be formalised by using a checklist to ensure that the trainee approaches all elements of the work correctly. It is prudent for departments to cover all major topics in this way and record the trainee's performance. In the United Kingdom the Royal College of Anaesthetists has published a curriculum in terms of competencies and departments are required to assess the trainee in some areas of practice at times that the College has stipulated.[3,4]

In the USA minimal requirements for experience are set (usually to be achieved in the first two years of training) by the ACGME (Accreditation Council for Graduate Medical Education http://www.ACGME.org). Yearly 'Experience logs' have to be submitted to this organisation. Clinical competency is assessed after each sub-specialty rotation and at least biannually (according to AGCME guidelines) and attested to by completion of a Clinical Competency Report that is sent to the American Board of Anesthesiology on a six monthly basis. The requirements also require minimal numbers of cases in sub-specialty areas. This would for example include at least twenty surgeries with cardio-pulmonary bypass and 40 anaesthetics for vaginal delivery. (See ACGME.org and follow the links: Program requirements, Anesthesiology, Educational Programme.)

The role of the consultant head of department/chairman

It is incompetent to allow an anaesthetic trainee who lacks the necessary skills, knowledge and professionalism to have the care of patients. It is the head of department's responsibility to ensure that this does not happen. It is therefore advisable to formally review the progress of new trainees and those new to the hospital and agree that they can undertake independent practice. A responsible teaching department will need some sort of framework for decisions about supervision and this framework will also determine how consultants and trainees come together for theatre teaching. The department may decide on further safeguards, such as insisting on

"Terrible! You always think somehow that it can never happen to you."

Advanced Life Support (ALS) or Advanced Trauma Life Support (ATLS) certification for doctors in some roles. In the US the AGCME insists that all residents have both basic and advanced life support accreditation. This must be renewed every two years.

The trainee's role in determining supervision

Trainees must accept much of the responsibility for the level of supervision they receive. They sometimes think that they can prejudice their career by appearing too cautious. This is not the case. On the other hand a gung-ho approach to work is always felt to be a serious flaw. Consultants should make it clear to trainees that they do not expect them to muddle through on their own and should explain to them the sort of circumstances in which they expect to be told what is going on. Kruger and

When discussing a case with a consultant:

- First – state the clinical problem clearly.
- Make sure the consultant has taken in the most important information.
- Tell the consultant what level of participation you would like from them.
- If you want to undertake the case with minimal supervision give your reasons for thinking you can cope.
- Tell the consultant if you have not met this sort of case previously.
- Don't accept an instruction to undertake a case with less supervision than you think is safe.

Dunning subjected a group of volunteers to tests of humour, logic and grammar.[5] They found that the least competent individuals had a tendency to overestimate their own performance. Their very incompetence interfered with their self-assessment. Although their test scores put them in the lowest quartile for performance they estimated themselves to be at the 62nd percentile. Paradoxically, improving the skills of the test subjects enabled them to be more aware of their deficiencies and their self-estimation reduced as their test score rose.

When a trainee telephones a supervising consultant they should first give a very short summary of the problem and explain why they are calling. At three a.m. it is not possible to make sense of a call that starts into the details of drug therapy and biochemistry before giving a clear description of the clinical problem. The trainee should indicate at an early stage whether they want on the spot help or telephone advice. The consultant cannot decide whether he or she needs to go but the trainee knows whether they need that help or not. Trainees should be taught to ask unambiguously for help if that is what they want. In addition the trainee should let the consultant know what they are feeling about the situation. Are they puzzled, worried, frightened or terrified? If they don't tell the consultant, then he or she will not know. When the telephone consultation is over both the trainee and consultant should ask themselves if they are happy about what has been decided. If the trainee is unhappy they must ring back and say so. If the consultant is unhappy they must get up and see the patient. The role of the trainee is to do the work of which they are capable and they must remember that until the consultant is by their side they are responsible for the patient. The safety of that patient must be their first concern. Part of the early training of anaesthetists must be about these responsibilities for the patient and how to ensure that they always get the help they need.

All trainees should, throughout their training, have many supervised lists in areas where they are already competent. This enables the consultant to watch the way work is organised and the trainee's manner in dealing with colleagues. It permits judgement to be made of the way the trainee applies his mind to routine work and whether vigilance is maintained in the face of familiar work. It also allows the trainer to watch the trainee deal with unexpected problems.

References

1. Hawkins P, Shohet R. *Supervision in the helping professions.* Philadelphia: Open University Press Milton Keynes, 1989.
2. *The CCST in anaesthesia I: General principles, a manual for trainees and trainers.* London: The Royal College of Anaesthetists, 2000;4.
3. *The CCST in anaesthesia II: Competency based senior house officer training and assessment, a manual for trainees and trainers*. London: The Royal College of Anaesthetists, 2000.
4. *The CCST in anaesthesia III: Competency based specialist registrar years 1 and 2, training and assessment.* London: The Royal College of Anaesthetists, 2002.

5. Kruger J, Dunning D. Unskilled and unaware of it: how difficult is in recognizing one's own incompetence lead to inflated self-assessments. *Journal of Personality and Social Psychology*. 1999;77:1121–1134.

Further reading

Morton-Cooper A, Palmer A. *Mentoring, perceptorship and clinical supervision. A guide to professional support roles in clinical practice.* 2nd ed. Oxford: Blackwell Science, 2000.

3

Learning from work

David Greaves

The curriculum for clinical teaching

Trainees need to know what to learn and consultants need to know what to teach. A syllabus is a list of things to be learned and a curriculum organises this into a scheme for teaching. Curricula for anaesthesia have been developed by individual training schools and by national organisations.* An integrated teaching scheme will need a curriculum and this may be in the form of learning objectives. The organisers will then decide how to satisfy each objective. They will need to decide whether clinical teaching, a lecture or some other sort of learning best meets it. Clinical teaching can satisfy many learning objectives wholly or in part. Clinical work should be arranged so that every trainee gets all the experience that is needed to meet the full range of practical objectives in the curriculum.

> *First, you must tell them what to learn.*
> *............Then you must tell them how you will test their learning.*

Competencies

A relatively recent introduction into medical postgraduate education has been defining curriculum as a list of 'competencies'. The competency based training movement

*Royal College of Anaesthetists, London, 1994, 1995, 1997, 2000; Royal College of Physicians and Surgeons of Canada, 1993; Australia and New Zealand College of Anaesthetists, 1992; American Board of Anaesthesiology, 1987 'Content Outline' Joint Council on In-Training Examinations, American Board of Anesthesiology, American Society of Anesthesiologists. Revised January 1996.

Competencies are tested by observing 'outcomes' — changed behaviour that can be attributed to the recent learning.

began in the USA in the mid-1960s when the quality of teaching in schools was criticised. The curriculum for teacher training was said to bear little relationship to the job requirements. With federal government sponsorship a number of colleges began to teach student teachers according to a new 'competency' based curriculum. This concentrated on skills for the classroom. Assessment became practice based and centred on observation of skills in action. Competency based training was introduced into education in the UK during the 1980s and has found a particular niche in vocational education. In a competency based scheme the professional task is divided into a list of the abilities of a 'competent' performer. Each item of the professional repertoire (such as putting in a cannula or listening to the heart) is a competency. Many competencies can be 'unbundled' into even smaller chunks.

Competencies may be items of knowledge, practical skills or ways of behaving. This is in line with the categories used to describe educational objectives. The competency-based scheme does not restrict the type of learning that can be used to reach the competency, nor does it restrict the time to learn the lesson. When the learner has mastered a competence it is possible to see that their approach to problems in this curriculum area is changed in predictable ways. These changes in behaviour are the 'outcomes' of learning and looking to find these changed behaviours assesses competency-based learning. Observation of competencies lends itself particularly to the teaching and testing of workplace skills and abilities.[1]

Observation of learning outcomes

A competency in the psychomotor domain:
'Performs cardiac compression effectively.'

A competency in the cognitive domain:
'Understands the causes of metabolic acidosis.'

A competency in the affective domain:
'Co-operates with the patient and midwifery team to negotiate a plan for pain management in labour.'

The testing of competencies looks at the 'outcomes of learning'. The competence model of the curriculum is linked to a new view of assessment. Traditionally a test is successful if it reliably returns a candidate's score in terms of a pass mark. In the new model the success of a test is defined in terms of how well a 'pass' guarantees future good performance. Traditionally the curriculum has been academic in nature and a major emphasis in testing has been on examinations. The examination hall has a reduced role in outcome testing and observation of the trainee in action becomes more important.[2]

The controversy over competencies in the professions

An argument has developed between those who see competence setting as useful and those who fear that a list of competencies does not capture the essence of clinical practice, because professional work is more than the sum of the relevant competencies.[3,4] Learning based on competencies and the testing of outcomes may not take into account the richness of the process of professional work. It is difficult to set objectives and outcomes for learning that ensure understanding of the complex process of clinical reasoning. On the other hand, practical anaesthesia is a good example of a trade that includes a number of practical skills and understandings that can be listed and relatively easily observed. It is possible to use the concept of learning objectives to set out the scheme of learning without fully accepting the testing of competencies that follows on. A theme that recurs when considering assessment is that the public and their political representatives are concerned about its rigour. Competency based curricula and testing appear to them to satisfy this rigour.

'Behaviour' and 'Attitude'

Doctors who organise training programmes recognise that, though 'failing' doctors are uncommon, their problems are often related to behaviour rather than ignorance or lack of skill. Though these affective elements of performance are not usually identified in a traditional syllabus, they are learned and can be incorporated into a curriculum based on competencies. These affective competencies have outcomes that are relatively easy to observe. Recording these provides objective data to feed

Some examples of failures in the affective domain:

- Unpunctual
- Unable to co-operate with colleagues.
- Irascible 'has a short fuse'
- Disorganised approach to work
- Unable to take the lead in the team
- Unable to play a subordinate role in the team
- Dishonest
- Unable to communicate with patients
- Unable to communicate with colleagues

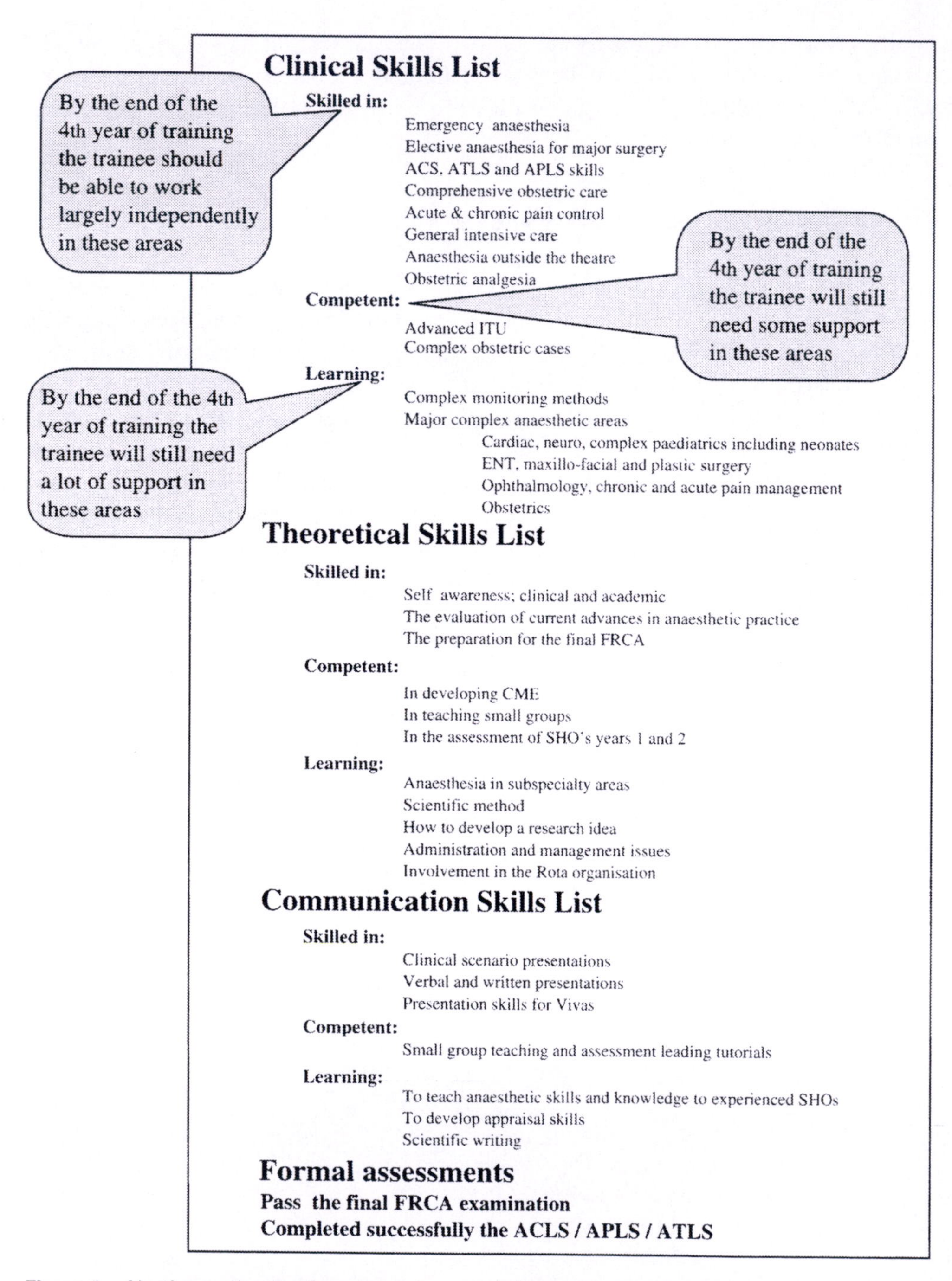

Figure 1. Northern schools of anaesthesia — overall competencies for SpR year 2.

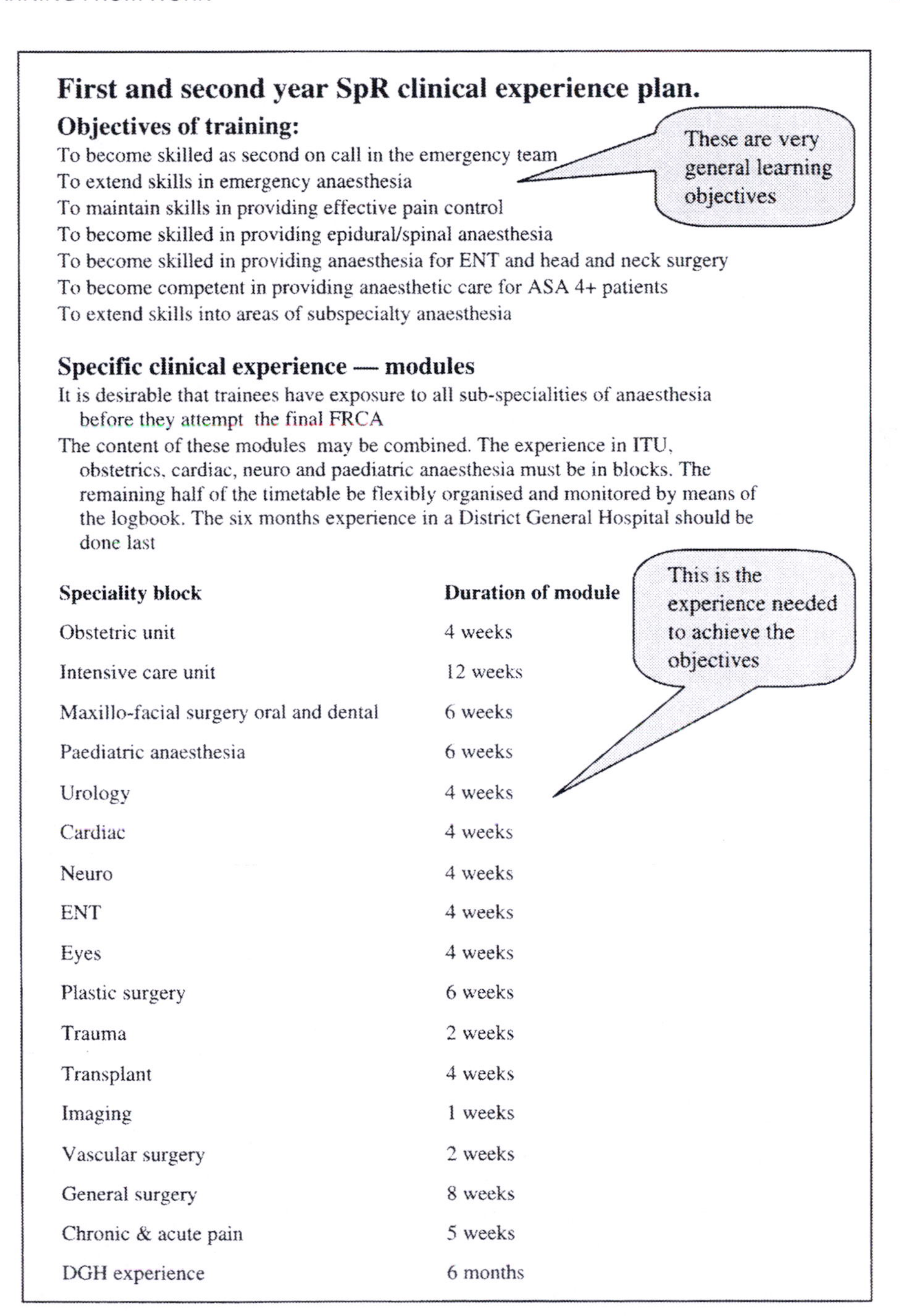

First and second year SpR clinical experience plan.

Objectives of training:

To become skilled as second on call in the emergency team
To extend skills in emergency anaesthesia
To maintain skills in providing effective pain control
To become skilled in providing epidural/spinal anaesthesia
To become skilled in providing anaesthesia for ENT and head and neck surgery
To become competent in providing anaesthetic care for ASA 4+ patients
To extend skills into areas of subspecialty anaesthesia

Specific clinical experience — modules

It is desirable that trainees have exposure to all sub-specialities of anaesthesia before they attempt the final FRCA

The content of these modules may be combined. The experience in ITU, obstetrics, cardiac, neuro and paediatric anaesthesia must be in blocks. The remaining half of the timetable be flexibly organised and monitored by means of the logbook. The six months experience in a District General Hospital should be done last

Speciality block	Duration of module
Obstetric unit	4 weeks
Intensive care unit	12 weeks
Maxillo-facial surgery oral and dental	6 weeks
Paediatric anaesthesia	6 weeks
Urology	4 weeks
Cardiac	4 weeks
Neuro	4 weeks
ENT	4 weeks
Eyes	4 weeks
Plastic surgery	6 weeks
Trauma	2 weeks
Transplant	4 weeks
Imaging	1 weeks
Vascular surgery	2 weeks
General surgery	8 weeks
Chronic & acute pain	5 weeks
DGH experience	6 months

Figure 2. Northern schools of anaesthesia — specific clinical experience needed to achieve the overall competencies.

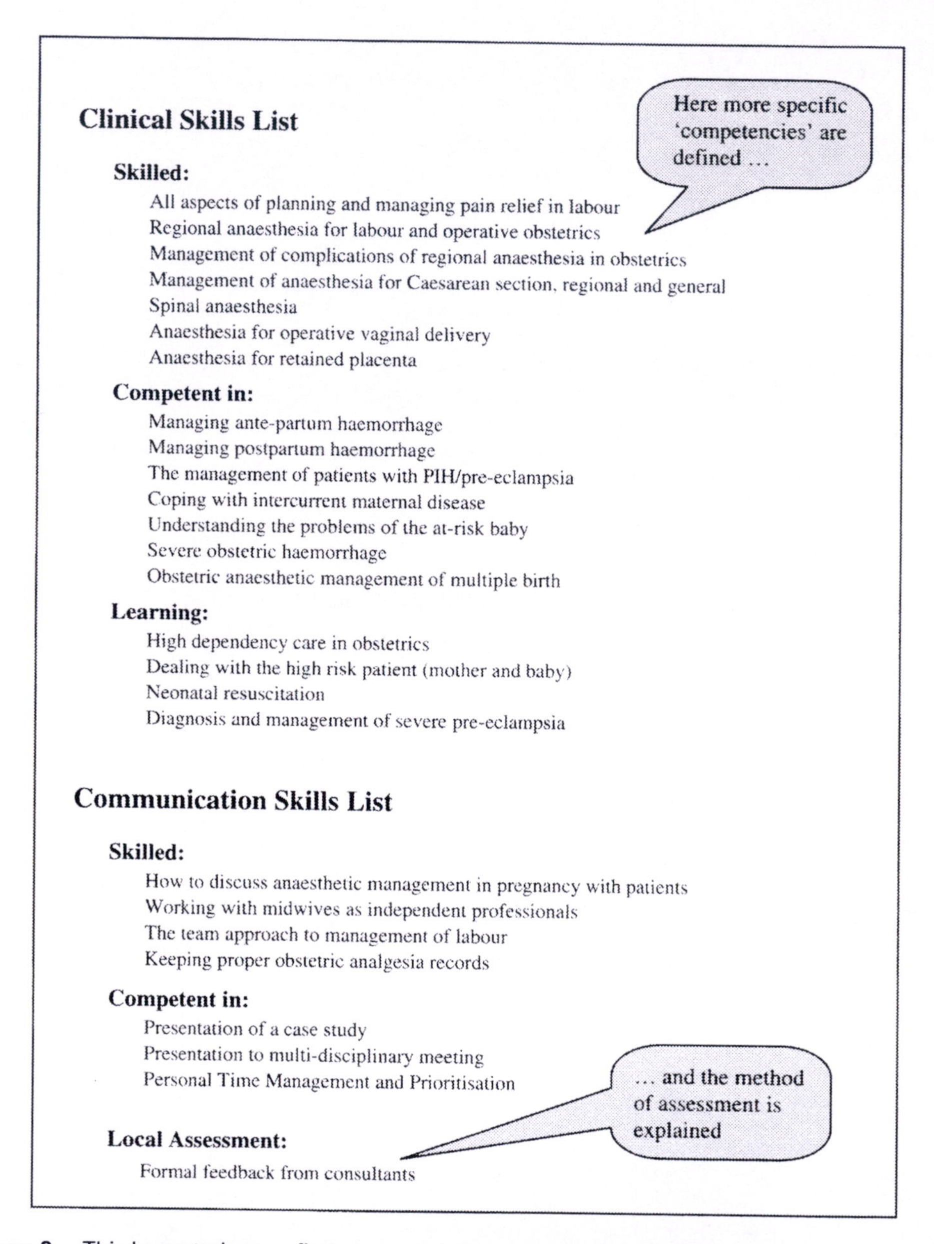

Clinical Skills List

Skilled:

All aspects of planning and managing pain relief in labour
Regional anaesthesia for labour and operative obstetrics
Management of complications of regional anaesthesia in obstetrics
Management of anaesthesia for Caesarean section, regional and general
Spinal anaesthesia
Anaesthesia for operative vaginal delivery
Anaesthesia for retained placenta

Competent in:

Managing ante-partum haemorrhage
Managing postpartum haemorrhage
The management of patients with PIH/pre-eclampsia
Coping with intercurrent maternal disease
Understanding the problems of the at-risk baby
Severe obstetric haemorrhage
Obstetric anaesthetic management of multiple birth

Learning:

High dependency care in obstetrics
Dealing with the high risk patient (mother and baby)
Neonatal resuscitation
Diagnosis and management of severe pre-eclampsia

Communication Skills List

Skilled:

How to discuss anaesthetic management in pregnancy with patients
Working with midwives as independent professionals
The team approach to management of labour
Keeping proper obstetric analgesia records

Competent in:

Presentation of a case study
Presentation to multi-disciplinary meeting
Personal Time Management and Prioritisation

Local Assessment:

Formal feedback from consultants

Figure 3. Third year trainee — first year specialist registrar obstetric learning plan.

back to the trainee, and ultimately provides the sort of evidence on which a doctors training can be stopped. When faced with a colleague who disrupts the clinical team by their behaviour many doctors will say, "There is a problem, but there is no question that he is clinically competent." It cannot be too strongly stated that it is incompetent to behave disruptively at work.

The same basic principles relate to teaching in the affective domain as in the domains of knowledge and skill. First, tell the trainee, in some detail, how they should conduct themselves professionally, then tell them what observations you will make to satisfy yourself that they are meeting the requirements.

Experiential curriculum

Most learning comes from experience. Learning can be organised on the back of experience, and such learning is called experiential learning.[5] Tweed and Donen proposed that an 'experiential curriculum' could be applied to the bulk of anaesthesia teaching. They believe that learning is a process that begins with experience and that what the learner does will therefore determine the outcome. In their view most learning can start with a clinical experience, and organising these experiences determines an experiential curriculum.[6]

It falls to College Tutors, Programme Directors and 'rota makers' to organise the work of trainees. Rotations and placements are usually organised on the basis of allocation to surgical specialties. In an experiential curriculum or competence based system consideration should be given to alternative methods of allocation, that are directly related to groups of competencies.

A lot of topics that anaesthetists have to learn are spread throughout a number of surgical specialties. This applies to monitoring techniques, for instance. A unit of work, organised to best learn about monitoring techniques, would include placements to see cardiac surgery, cardiac investigations, arterial surgery, intensive care and neurosurgery. It might also include visits to the medical physics and engineering departments. Many lessons in anaesthesia such as: acute pain control, pre-operative anxiety management and anaesthesia for patients with heart disease, would best be learned by experience that is organised in new ways. Clearly surgery is central to the insults heaped upon our patients, but the 'surgery-centred' allocation of work leads to a similar approach to anaesthesia.

Encouraging role models

If role models and an apprenticeship pattern of work are important in medicine, and there is reason to believe that they are, then we should consider ways of allocating experience that will maximise these. Trainees could form "firms" with consultants. In some cases these might reflect sub-specialty divisions and give intensive experience in areas that need continuity of learning (obstetrics, pain, intensive care). But in many instances, nothing is lost by attaching the trainee to the variety of lists undertaken by two or three consultants. This would provide a work pattern conducive to apprenticeship and in-service assessment. A small group of consultants would be responsible for the trainee's educational well-being.

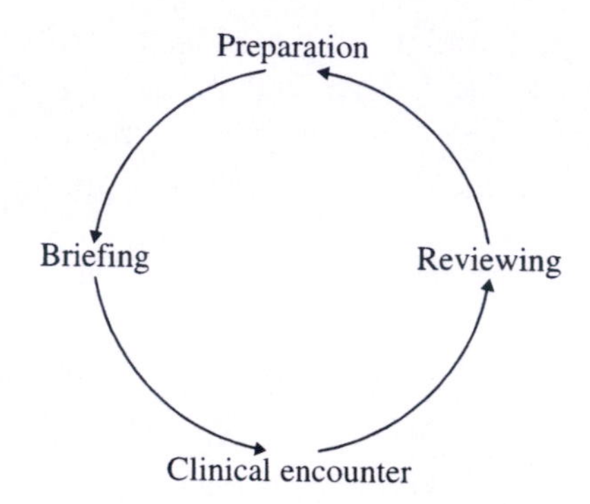

Figure 4. The experience cycle.

Making sense of clinical teaching — review, reflection and explication

Learning builds on the scaffold of existing knowledge that has to be adjusted with each addition.

Cox has described the cycle of clinical learning (see Fig. 4).[7]

- Preparation for teaching comes first. Existing knowledge and beliefs are organised for the new teaching.
- Introduction to the clinical teaching is the process of briefing.
- There follows the clinical encounter.
- Review follows the clinical teaching during which the teacher draws together the strands that have emerged from the passage of learning.

During the working day, with a list of patients, the experience cycle may be repeated many times. Whilst the experience cycle describes the components of clinical teaching the process of learning from experience is more complicated, and takes place after the lesson. Cox described a second cycle of learning that he termed the explanation cycle

In the USA the competency movement has culminated in the definition of six core competencies for medical practitioners. These are:

- Patient care
- Medical knowledge
- Practice based learning and improvement
- Inter-personal and communication skills
- Professionalism
- Systems based practice

These competencies must be expanded and defined appropriately for each specialty.

Information about this is available at www.AGCME.org

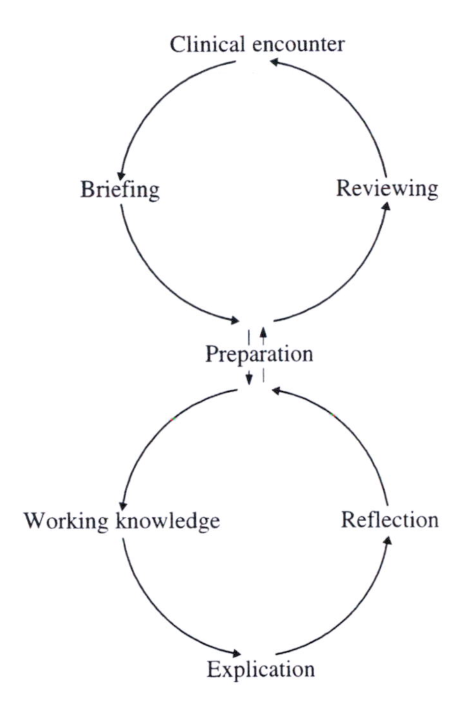

Figure 5. The explanation cycle.

(see Fig. 5). A significant difference exists between what he termed review, occurring with or soon after the events, and reflection, a later process. Explanation follows at the end of the experience cycle and is the process that makes the trainee approach his or her next lesson with the outcome of the preceding lesson having modified their underlying experience, and hence their state of preparedness for learning. It should not be taken for granted that this process will occur.[8]

Reflection

Active experience is needed for clinical learning, but it is not sufficient in itself. It must be followed by review; a visible external process, which then leads on to reflection on the invisible, internal meaning of the experience. Reflection is the process of understanding experiences and looking for deeper meanings.

Reflection leads to explication of the experiences. At this stage of learning the trainee should be left to explore his thoughts, without the consultant interfering. It is important that the learner seeks out the further questions he wants to ask for himself, because the consultant does not know what the trainee already believes. A new piece of information, that does not fit with the existing system of belief, may be the lead that prompts the trainee to realise that his previous understanding might be flawed.

Reflection is followed by explication. Explication is the process of re-erecting the scaffold of belief to incorporate the new learning. It can be a straightforward revision of ideas and addition of the new, or it can be a complex process. The trainee may need to revise and review areas of knowledge, and he may also need to study privately in order to tidy up knowledge. The new learning may have indicated that knowledge is incomplete in a number of areas. Books, journals and practical learning may be needed to complete the explication process. Explication has resulted in the development of a new working knowledge. The consultant who has the good fortune to work with the same trainee on a regular basis can help with the transition from reflection to explication by careful enquiry into the trainees new decision framework. Informal questions may be asked such as, 'How has last weeks experience affected the way you will approach problems?' Alternatively, the trainee can be asked to draw a decision tree or algorithm relating to a problem, and the consultant can then look for evidence that the scheme reflects the recent experience.

Explication leads back to preparation. When the trainee has incorporated his new learning, and has re-adjusted his system of beliefs, the reflective process will probably have revealed loose ends that require to be tidied up. This is the process of

> Doctors who behave inappropriately in the workplace are clinically incompetent.
>
> Reflection is the key to learning:
> The habits of reflective practice should be learned at the start of a career and continue throughout it.
>
> Hints for trainees:
> Read about and understand the process of critical reflection.
> Cultivate the habit of thinking over problems you have met—after the events.

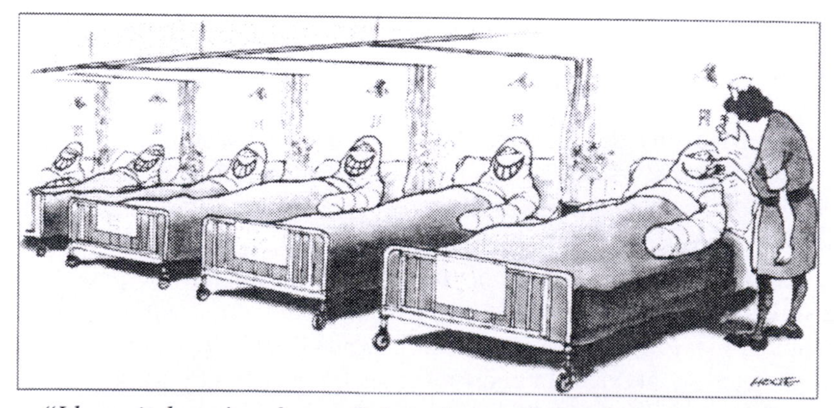

"I know it doesn't make you feel any better, Mr Pendleton, but it makes my job infinitely more bearable."

getting ready to learn. The trainee has questions to resolve and clinical learning activities to pursue. The trainee is now prepared for a new step in learning.

It might be thought that this formal approach to the resolution of a teaching experience would only be needed when some major event is bringing about a major shift in the trainee's belief. Trainers can know little of what their trainees believe, and an apparently successful trainee can have a misconception at the root of his knowledge. In this case, learning a relatively mundane fact can precipitate a major upheaval. It is important to remember that the mind has no difficulty in holding irreconcilable views simultaneously, and of regularly making a paradoxical decision — if such behaviour is associated with successful outcomes.

Clinical teachers should understand these complex patterns of learning, and organise their sessions in ways that facilitate review, reflection and explication. In particular they must give the trainee the opportunity to return to a topic after reflection. The reflective period may be days, weeks or even months.

References

1. *The CCST in anaesthesia II: Competency based senior house officer training and assessment, a manual for trainees and trainers*. London: The Royal College of Anaesthetists, 2000.
2. Wolf A. *Assessing assessment. Competence based assessment*. Buckingham Philadelphia: Open University Press, 1995;30–52.
3. Myerson KR. Can we assess professional behaviour in anaesthetists? (Editorial.) *Anaesthesia*. 1998;53:1039–1040.
4. Greaves, JD. Anaesthesia and the competence revolution. *British Journal of Anaesthesia*. 1997;79:555–557.
5. Kolb DA. Experiential learning. Englewood Cliffs, New Jersey: Prentice Hall, 1984;20–38.
6. Tweed WA, Donen N. The experiential curriculum: an alternate model for anesthesia education. *Canadian Journal of Anaesthesia*. 1994;41:1227–1233.
7. Cox K. Planning bedside teaching. *Medical Journal of Australia* 1991;158:280–282, 355–357, 417–418, 493–495, 571–572, 607–608, 789–790;159:64–68.
8. Boud D, Keogh R, Walker D. *Reflection: Turning experience into learning*. London: Kogan Page, 1985;18–41.

Further reading

Boud D, Cohen Ruth, Walker D. *Using experience for learning*. Buckingham: The Society for Research in Higher Education & The Open University Press, 1993.

Schon, DA. *Educating the reflective practitioner*. San Francisco: Jossey-Bass, 1987;3–40.

Warner-Weil S, McGill I. *Making sense of experiential learning: diversity in theory and practice*. Buckingham, UK: Open University Press, 1989.

4

The assessment of clinical competence

David Greaves

To plan learning, monitor progress and keep patients safe consultants need to estimate the clinical competence of trainees. There are rigorous, well researched methods available for testing knowledge, but assessments of clinical performance are often neither structured nor standardised.

What is clinical competence?

It is difficult to define clinical competence.[1] 'Adequate clinical proficiency, as judged against the performance of an accepted specialist', is one way to define competence. Performance can be assessed for each component of the work e.g. was the procedure of inserting the arterial line satisfactory? Clinical competence is, however, a judgement of the whole clinical task, and incorporates knowledge, practical skills, clinical behaviour, past experience and personality. These elements together permit a standard of performance that is judged satisfactory by those who have an interest in the outcome. The doctor, the patient, the doctor's colleagues and society as a whole all make this judgement, and in doing so they lend particular weight to the aspects of whole performance that seem important to *them*. No wonder a simple definition of 'competent' is hard to arrive at.

Incompetence can be due to problems with part of the professional task. A doctor with excellent knowledge and practical abilities can still be incompetent, if some other factor in performance, such as the way he behaves, results in an unacceptable standard of care. Everybody can make mistakes; how often does performance have to be poor for the practitioner to be judged incompetent? Habitual poor performance betokens incompetence, but then so do sporadic lapses, that can be linked to a particular deficiency in the doctor's clinical method. Omission, such as not attending to see a patient and commission, such as walking away from a clinical situation due to anger, are also to be categorised as incompetence, even though the doctor knew what to do, and could have skilfully dealt with the problem.

Assessment

An educational assessment measures performance against a standard.[2] As learning progresses performance will be better. Assessment serves two purposes:

Educational measurement must be:
- Fair
- Valid
- Reliable
- Practicable
- Appropriate in its effect on learners
- Credible

- Monitoring learning
  Assessment can give information about a learners current performance, so that they know how well they are doing and in order to plan their learning. This is called formative assessment.
- Setting standards
  Assessment can be against a formal, agreed standard. Marks can be used to rank examinees, or an agreed minimum performance can be used as a pass mark. Public and Political considerations often dictate that the mechanisms of summative assessment are formal, seen to be impartial and rigorous. This may lead to the choice of a test for its perceived rigour, even when it is not the best way to assess the abilities in question.

Standards used for assessment:
- *Peer referenced*: used to rank a group of learners by comparing their scores.
- *Ipsative assessment*: used to measure progress against the learners previous standard. Used to demonstrate progress.
- *Criterion referenced*: used to measure performance against an agreed standard e.g. derived from the consensus of examiners.
- *Limen referenced*: used when the standard is measured only on a pass/fail scale. Performance is either satisfactory or unsatisfactory.

The qualities of an assessment

Fairness

Assessment must be fair. The candidate must have been correctly informed about the content of the test and how it will be conducted. If the outcome is competitive, no candidates should have privileged knowledge of the questions. No candidate should experience prejudice because of who they are, who their supporters are or where they are working.

Fairness requires that the consultants making assessments must have the necessary skill to make the assessment. A few doctors who are thought to be experts, compose questions for written examinations, and a small group of examiners can design and conduct a written test by MCQs. For in-service clinical assessment many consultants must take part. Those who organise a group of assessments must ensure that the individual assessors have been formally trained how to make the observations. Candidates that fail will often believe that the test was unfair, because the examiner allowed irrelevant issues, or personal prejudice to cloud their judgement. Fairness requires that the opinion of several consultants go into the overall grade setting. Even then, candidates may believe that one examiner has influenced the others. Great care must be taken to make assessment transparent as well as fair. Fair in-service assessment is largely the responsibility of the trainers who administer it. If a decision based on an in-service assessment is challenged in court, it will probably be on the basis that it was unfair.

In the case of in-service assessments, the assessment must not be anonymous. All comments about behaviour or untoward events in the conduct of anaesthesia must include details of time, date and witnesses. Ideally, supervising consultants should challenge the trainee with the problem at the time, and give them the opportunity to justify themselves. This is a function of giving feedback.

Types of validity:

- Face validity
- Content validity
- Predictive validity
- Concurrent validity
- Construct validity
- Criterion validity

Validity

A valid test is one that tests what it is supposed to measure. There are a number of separate aspects to validity.

Face validity is a measure of how appropriate the test items are to the purpose of the assessment. If I want to test a person's performance in recognising birds, an appropriate test would include pictures of birds, rather than squirrels.

Content validity is the extent to which a test is representative of the full range of knowledge or skills that have been learned. My test of bird recognition should include a suitable sample of all types of birds.

Predictive validity indicates whether a test result is a good predictor of future performance. Is winning academic prizes at high school correlated with good performance at medical school?

Concurrent validity is a measure of how well the test correlates with a test that is believed to test the same skill. How does a multiple choice question test correlate with a short answer test of knowledge of the pharmacology of muscle relaxants?

Construct validity relates to the 'construct' of the skill being tested. In this sense construct implies the idea or conception of the skill in the real world. Construct validity is a judgement of the adequacy and appropriateness of the inferences about the subjects, that are being derived from the test scores. It relates the test result to real behaviour. Are the measurements appropriate for the category of learners who are being tested, and for the ultimate use to which they will put the capabilities that are being measured? Anaesthetists have knowledge, skills and attitudes that are particular to their profession, and they use them in ways that reflect their specific professional role. A test with high construct validity for anaesthetists would be one in which proficient anaesthetists, expert in the work, always scored well, whereas, anaesthetists who show lapses in the real work would score badly.

I have dwelt upon the meaning of construct validity, because it is fundamental if assessments of whole work are to be made, either in the real workplace or in simulation. It is commonly assumed that performance in a simulator will reflect the subject's performance in real work. This assumption may be made if skilled anaesthetists, universally upheld as superior performers, perform well in simulation, and known poor performers do badly. This aspect of a test needs to be investigated formally. Workers sometimes assert good construct validity for a test, because scores increase when more experienced individuals are tested. It cannot necessarily be assumed that seniority brings proficiency in all aspects of work.

Criterion validity relates the candidate's performance to established standards of practice. Is the paper too hard or too easy? Are the test items selected to have the degree of difficulty that is appropriate for the standard of the learners?

Perhaps a more useful concept is to look at the overall validity of a test, and consider aspects of each category in a single assessment. 'Construct' is the overall theme of this approach and emphasis is placed on the way the test is used, rather than the way it was designed. A test that satisfies all criteria can be mis-used and rendered invalid.[3] A unified scheme developed by Mesnick for investigating a test is shown in the box.

Questions to ask about the validity of an assessment:[3]

- Are we looking at the right things in the right balance?
- Has anything important been left out?
- Does our way of looking introduce sources of invalidity or irrelevant variance that bias the scores or judgements?
- Does our way of scoring reflect the manner in which domain processes combine to produce effects and is our score structure consistent with the structure of the domain about which inferences are to be drawn or predictions made?
- What evidence is there that our scores mean what we interpret them to mean? In particular, as reflections of knowledge and skill having plausible implications for educational action relative to personal and group standards.
- Are there plausible rival interpretations of score meaning or alternative implications for action, and if so, by what evidence and arguments are they discounted?
- Are the judgements or scores reliable and are their properties and relationships generalisable across the contents and contexts of use as well as across pertinent population groups?
- Do the scores have utility for the proposed purposes in the proposed settings?
- Are the scores applied fairly for these purposes?
- Are the short- and long-term consequences of score interpretation and use supportive of the general testing aims and are there any adverse side-effects?

Reliability

Was the trainee's pass in the exam a fluke? Was their fail unexpected? The reliability of a test measures the consistency and precision of the observations — its' accuracy. A test should be stable enough to give the same result if applied more than once within

Reliability — is the test accurate?

- Does the result change on a retest?
- Does the result vary between examiners?

a few days, and should not vary between examiners. There should also not be much variation between an assessment and a repeat assessment with similar test items.

It is easier to investigate the reliability of conventional tests than their validity. In the case of high-stakes tests, such as medical graduation, society does not want incompetent doctors to get lucky on the day, so a lot of attention is paid to reliability.

Reliability is improved if a variety of types of tests are applied, by a variety of examiners, on many occasions. In medical assessment, however, concern about a chance pass by a poor candidate tends to lead to overemphasis on individual parts of the exam.

Practicability

It's not much good designing the perfect test if you cannot apply it.

All methods of measuring clinical performance are time consuming and expensive. Some types of test, such as the Multiple Choice Question Test, are easy to apply, mark and standardise. This tempts examiners into using such a test in circumstances where the MCQ is not a good test for the aspect of performance under test. The effort of administration, interpretation and feedback for in service assessment is formidable. Criteria for scoring have to be developed, and consultants will have to be trained in how to apply them.

Appropriate effect on learners

Setting an assessment modifies the learning behaviour of trainees.

Learners always direct their learning towards tests, so it is important that these will encourage useful learning. Examinations may have content that is of little use to the candidates in their work place. Learning this material will distract them from their workplace-based learning. A topical example of the possible malign effects of examination content on examinees would be the use of MCQs from the current training schemes to assess specialists for re-accreditation. Consultants often do not remember these basic sciences, and probably do not need to. If they are tested in these areas they will concentrate their continuing education efforts into passing, and will probably neglect education that could move forwards their standard of practice.

Credibility

It is increasingly necessary to consider the credibility of tests in medicine. Certification and re-certification tests permit doctors to do dangerous things to patients. These patients, as a whole, are required to trust their doctors. Part of this trust comes from knowing that appropriate methods have been used to weed out poor performers. Such tests must be valid and reliable, and it is easier to show that some tests, such as Multiple Choice Questions, meet the criteria. The battery of tests used for certification must be well enough researched and documented to inspire confidence in the public and their representatives. At the same time this must not lead to the use of inappropriate but valid and reliable tests, at the expense of appropriate tests, the efficacy of which is more difficult to document.

What do we want from the assessment of clinical competence?

Each of the parties to an assessment has their own expectation from that assessment. It is important to understand this, as their interpretation of what is competent will reflect their standpoint. The instruments chosen for an assessment will reflect a compromise between these various interests.

The trainee
A trainee needs to learn the trade and pass the accreditation exams. Trainees see examinations as a hurdle. They need to feel that tests are fair and not unnecessarily difficult. They would prefer assessments that are relevant to their work and would probably welcome assessment of their performance in theatre. They usually have clinical skills of a high order and would like an assessment that gives credit for this.

Training directors
Training directors are concerned with developing clinically competent anaesthetists. Training directors also need trainee assessment that will give good information about the effectiveness of the teaching and clinical experience that they are organising.

Clinical teachers
The consultants who supervise clinical work teach practical anaesthesia. Their concern is with safety and the clinical proficiency of the trainee. They need to know what supervision the trainee needs and what tasks he or she can do safely. Clinical teachers also need information on how effective they are as teachers.

The hospital manager
All apprenticeship systems allow increasing responsibility with decreasing supervision. Managers require reassurance that supervision of trainees is always matched to their performance.

Accrediting organisations
It is usual for specialist status to be dependent on passing an assessment organised by a professional organisation. Such specialist organisations will demand high academic standards and will favour assessment systems that have high status and are easy to administer to large groups of trainees.

The general public
The general public wants doctors to be competent. They believe that assessment guarantees this and, in general, have faith in large-scale external examinations. Society would like to have assurances that doctors can perform properly it is probable that the public would be surprised to find that doctors are not subject to a 'driving test'.

What do we want to measure when we assess clinical competence?

When we assess clinical competence we try to obtain a measure of performance that applies to the whole clinical task. This will be used to target special training on trainees who do not cope well. Ultimately it will be the basis on which trainees with persistent problems are refused specialist accreditation.

In undergraduate education, tests of clinical proficiency are commonplace, and the means for applying them and standardising them are well developed. Such tests are, however, very restricted in scope. It is easy to test the ability to hear and interpret heart sounds or perform sequences of cardiac massage. Using standardised patients it is possible to evaluate proficiency with specific clinical problems involving clinical history taking and examination. Such tests and simulations can equally be applied to postgraduate education.

To test competence, however, we would really like to know how trainees behave in real situations, how they make decisions and whether their judgements can be relied upon. Real situations are more complex than simulations or exercise of single skills. They are also unpredictable, and are framed in a different environment: the workplace rather than the examination hall.

Setting standards for measurements of clinical performance

At present the clinical competence of trainees is often based upon the reports of their trainers. Such assessments are called In-Training Assessments (ITA), and are based on the subjective professional judgements of supervising consultants. The development and use of specific criteria is discussed in Chapter 16. Do trainees accept the use of subjective data in their accreditation? Chandra et al. attempted to develop weightings for using what was previously a formative examination in pharmacy, as a summative assessment that counted towards certification.[4] The group of students felt that they should undergo frequent, formative assessment and that this should be given adequate weighting in the final assessment grading. A good correlation was found between the marks in the formative and certifying examination.

Rhoton et al. attempted to relate non-cognitive skills to performance, as estimated by the occurrence of near misses in anaesthetic practice.[5] In a large study across five teaching hospitals, they analysed the comments (9000) made by faculty, relating to 45 residents. Daily CASE (Clinical Anaesthesia System of Evaluation) forms were examined. These forms record faculty comments in a number of domains, according to a rating scale. Non-cognitive skills that were predictors of performance were conscientiousness, composure, management ability, confidence, crisis management and knowledge. Scores for conscientiousness and composure predicted two thirds of the variability of critical incidents.

Self-assessment (from Boud[6]).

- It is beneficial to establishing a pattern for life-long learning.
- It needs to be developed during higher education.
- The skills of self-assessment are not developed at school or medical school.
- The skills of self-assessment may not be fully transferable between domains.
- It is necessary for effective learning.
- For effective learning to take place learners must develop the capability of monitoring their own learning strategies.
- Self-assessment throughout the course of learning develops the skills of critical reflection.
- Self-assessment is reliable only if the learners are taught how to do it and provided with a framework to work within.

Self-assessment

Where assessment is seen to be a test of competence patients have an interest in the rigour of the test. Self-assessment is distrusted by the public and politicians who feel that trainees will be soft on themselves. It is hard to explain why learners would fail themselves, particularly when this will disadvantage them.

Self-assessment is usually concerned primarily with students developing their learning skills. As such, the assessment is a form of reflection on learning. All assessment involves two tasks: establishing criteria for performance and judging work against them. Self-assessment involves the learners in both these processes. Learners are often involved in only the latter half of this process, and this limited function can be described as self-grading. True self-assessment also includes the involvement of the learner in deciding what good work is. In a professional sphere this often includes the interpretation of existing guidelines on what constitutes good practice. Self-assessment has been championed as having a number of advantages for learning. Boud has enumerated a number of premises on which the practice of self-assessment may be based.[6]

There has been quite a lot of experience of self-assessment in medicine. A typical report from the early 1990s was of a longitudinal study of medical students followed through two years of medical school.[7] There was no correlation between the students-self-assessment and the scores they received. Gordon reviewed the literature on the validity of self-assessment in medicine and found support for the view that the validity of self-assessment is low, but that the capacity of learners to self-assess only improves a small amount with practice.[8] In five programs where explicit self-assessment goals and training strategies were developed, the validity of self-assessment seemed much better. In a subsequent review, Gordon examined 11 studies that met his criteria for teaching the students how to self-assess, and found that they demonstrated satisfactory levels of validity.[9]

In implementing self-assessment a number of non-cognitive benefits such as improved morale and motivation are seen, along with cognitive benefit in terms of improved performance. Sigiura et al. used a rating scale for the assessment of trainee anaesthetists.[10] They found good agreement between the scores by self assessment and by the supervising doctor. Indeed the supervisor scores were higher by one level out of five. In a study of medical students undertaking a third year term in anaesthesia, who completed self-assessments, there was very poor correlation between the students assessments and those of the consultant trainers.[11] A more recent study in surgical practice showed better examination scores for trainees who made major use of a teaching system that involved self assessment when compared to those using other learning methods.[12] In another recent study of medical students, their interviewing skills were assessed by videotape.[13] These workers found a high correlation of trainee and supervisors assessment that they attributed to the training in self-assessment that the learners had received. Other reports that have supported the validity of self-assessment in medicine have been those of Fincher, Lewis and Kuske and Fincher and Lewis.[14,15]

It seems that self-assessment with real summative functions would benefit learning, but politicians are not ready to understand that it is not a soft option.

References

1. Hager P, Gonczi A. What is competence? *Medical Teacher*. 1996;18:15–18.
2. Grant J, Jolly B. (eds). *The good assessment guide*. London: Joint Centre for Education in Medicine, 1997;10–21.
3. Mesnick S. Validity. In: Linn RL, editor. *Educational measurement*. 3rd ed. New York: Macmillan, 1989;13–103.
4. Chandra D, Kulshrestha S, Chandra M. Weightage of formative examinations in certifying examination in pharmacology: an opinion poll and relative performance of learners. *Medical Teacher*. 1992;14:197–200.
5. Rhoton MF, Barnes A, Flashburg M, Ronai A, Springman S. Influence of anesthesiology residents' noncognitive skills on the occurrence of critical incidents and the residents' overall clinical performance. *Academic Medicine*. 1991;66:359–361.
6. Boud D. Enhancing learning through self assessment. London: Koogan Page, 1995; 1–33.
7. Frye AW, Richards BF, Philp EB, Philp JR. Is it worth it? A look at the cost and benefits of an OSCE for second year medical students. *Medical Teacher.* 1989;11:291–293.
8. Gordon MJ. Self-assessment programs and their implications for health professions training. *Academic Medicine.* 1992;67:672–679.
9. Gordon MJ. A review of the validity of and accuracy of self assessments in health professionals training. *Academic Medicine*. 1991;66:762–769.
10. Sugiura Y, Miyamoto E, Harada J, Goto Y, Takahashi K. Evaluation of the training of the anesthesiologist. [In Japanese]. *Masui*. 1996;45:766–768.

11. Sclabassi SE, Woelfel SK. Development of self-assessment skills in medical students. *Medical Education*. 1984;18:226–231.
12. Wade TP, Kaminski DL. Comparative evaluation methods in surgical resident education. *Archives of Surgery*. 1995;130:83–87.
13. Farnhill D, Hayes SC, Todisco J. Interviewing skills: Self-evaluation by medical students. *Medical Education*. 1997;31:122–127.
14. Fincher R-ME, Lewis LA, Kuske TT. Relationships of interns' performances to their self-assessments of their preparedness for internship and to their academic performances in medical school. *Academic Medicine*. 1993;68:S47–50.
15. Fincher R-ME, Lewis LA. Learning, experience, and self assessment of competence of third year medical students in performing bedside procedures. *Academic Medicine*. 1994;69:291–295.

Further reading

Gipps CV. *Beyond testing*. London: The Falmer Press, 1994.
Grant J, Jolly B. (eds). *The good assessment guide*. London: Joint Centre for Education in Medicine, 1997;10–21.
The metric of education. *Medical Education*. 2002;36:9 (special issue on assessment).
Wolf A. *Assessing assessment: Competence-based assessment*. Buckingham: Open University Press, 1995.

5

The ethics of learning on patients

Catherine Bartley

> *"I will impart this Art by precept, by lecture and by every mode of teaching..."*
> The Hippocratic Oath.

A subject that often arises at meetings where specialists discuss clinical teaching is whether it is reasonable to teach a trainee to perform clinical work, knowing that you could do it better yourself. Does a patient have the right to treatment by an accredited specialist? One of the oldest arguments between ethicists is whether a deed is

"In my opinion he's got an unfortunate manner."

ethical if it accomplishes the 'most good', i.e. 'the end justifies the means', as claimed by Jeremy Bentham and John Stuart Mill, or whether, as maintained by Immanuel Kant, there are principles of morality which cannot be ignored, regardless of their consequences. Examples of such principles may include beneficence, non-malfeasance, respect for autonomy and distributive justice, but many other moral codes exist.

At first glance the welfare of the patient and the need for junior trainees to learn practical skills 'on real people' seem to be irreconcilably in conflict. It is obviously essential for there to be a supply of well-trained anaesthetists for the future welfare of patients; this is a worthy 'end' to aim for. However, this is an example of the conflict discussed above; in seeking this end, the doctor may be compromising other principles of medical ethics.

Should a consultant ask a patient's permission for a trainee to undertake anaesthetic procedures? Everyone has to learn.

Is occasional underperformance due to inexperience acceptable as the price of having a skilled, well-trained medical work force?

Is it appropriate to compromise the welfare and safety of the individual patient, for the good of the population as a whole?

The patient who is a subject for training may be viewed as deriving no benefit himself, and even risk some harm, by agreeing to allow an inexperienced doctor to learn a procedure 'on him'. Does the fact that this is for the 'greater good' justify breaching that primary doctrine of medicine, 'First do no harm?'

One area where this issue has been studied in detail is non-therapeutic clinical research. Here too, the subject derives no personal health benefit from participating in the research, but it is considered acceptable to perform such research, as long as certain conditions to safeguard the well-being of participants are fulfilled. These conditions are laid down in the Declaration of Helsinki, and could be adapted to apply to training doctors on patients:

Considerations when training junior anaesthetists on patients (adapted from the 1989 revised Declaration of Helsinki)[1]

1. *The purpose of training a doctor using human subjects must be to improve that doctor's clinical practice.*
2. *Only appropriately trained, competent and qualified practitioners must conduct such training. These practitioners should be competent at performing the procedure being taught, and be competent trainers too.*
3. *Training involving human subjects cannot be legitimately carried out unless the importance of that specific trainee learning the technique is in proportion to the inherent risk to the subject.*

This condition implies that when teaching more 'risky' techniques to trainees, one must balance the risk to the patient against the value to society of this particular trainee learning this particular technique. If a procedure is performed relatively infrequently, then the 'patient resource' should be used appropriately to 'maximise the good' for society, and similarly it seems unjustifiable to expose a patient to the increased risk of being a subject for training, especially if the technique has a significant morbidity and mortality, if the trainee will never gain enough experience to become competent at that technique.

4. *Every procedure to be taught using a patient should be preceded by careful assessment of predictable risks in comparison with foreseeable benefits to the subject or to others. Concern for the interests of the subject must always prevail over all other concerns.*

It will usually be inappropriate to perform a procedure on a patient purely for training purposes: the procedure must also be of benefit to the patient.

5. *The right of the patient to 'safeguard his or her integrity' must always be respected. Every precaution must be taken to respect the privacy of the subject and to minimize the impact of being used in a training exercise on the patient's physical and mental integrity and on the personality of the subject.*

> *The need to teach or learn is not sufficient reason to change a patient's treatment, e.g., if you don't think an arterial line is clinically indicated don't do one to let the trainee practice.*

This rather vague right of the patient to 'integrity' in the Helsinki Declaration can be seen as a patient's right to privacy, respect and honesty from others.

6. *Doctors should abstain from engaging in training involving procedures on patients unless they are satisfied that the hazards involved are believed to be predictable. Doctors should not let a trainee learn a technique on a patient if the hazards are found to outweigh the potential benefits.*

The trainer must be sure that the trainee is capable of performing the procedure safely. This implies a need to assess each trainee's competence, knowledge and abilities, before allowing him or her to train on patients. If the procedure is not proceeding as it should and risks to the patient increase to an unacceptable level as a result of the training exercise, then the trainer must intervene.

7. *Record-keeping, critical incident reporting and assessment must be accurate.*
8. *In any training exercise on patients, each subject must be adequately informed of the aims, methods, anticipated benefits and potential hazards of the procedure and the discomfort it may entail. He or she should be informed that they can abstain from participation in the training exercise at any time. Ideally this consent should be documented.*

Ask yourself:

- Is the procedure necessary?
- Can the trainee do the procedure properly?
- Is the supervisor able to ensure that a lapse by the trainee will not result in damage to the patient?
- Is the patient aware of any risk that the learner represents?

There is the view that when a patient presents for surgery it is much like booking an air flight: the customer specifies the destination and the airline, but the specifics of what make of plane, who pilots the plane and his experience, who are the cabin crew etc. are left to the airline. One is never told if the pilot is landing a plane for the first time 'for real', and we trust that such novice pilots are adequately prepared for, and supervised in their job. Or, there is the less paternalistic view that patients should be given as much information and as many choices as possible, in order to respect their autonomy and to share responsibility for decisions between the patient and the medical team — like selecting the components for a new 'hi-fi'. Obviously some patients would prefer to buy a pre-assembled stereo system, but others are keen to weigh the pros and cons of every possible specification. Whether the anaesthetist views his anaesthetic as a 'package' or as a series of options for the patient will determine whether he feels it necessary to seek specific consents for different components of the anaesthetic. At present both approaches are common practice, and as such would be defensible in court.

The definition of 'informed consent' and the information felt necessary to give to the patient, is a matter of professional judgement for the doctor. For example, it seems ludicrous to gain specific consent for an experienced trainee to perform 'routine' anaesthetic procedures, which he has performed competently on numerous previous occasions, but does the patient not have the right to know if an epidural is the first one a trainee has ever inserted, and to request a more experienced operator? Evidence from recent high profile inquiries* showed that patients and their relatives were most upset at "being kept in the dark" over issues they felt were important. Many said that if their consent had been sought for the organs of their dead relatives to be retained for educational purposes, then they would gladly have consented, but the fact that they were not even asked was a serious breach of their autonomy.

A Norwegian survey showed that 69% of people would be happy for medical students and trainees to learn ventilation and airway skills on them whilst anaesthetised and those, who refused consent or were uncertain, were concerned about injury or just found the idea 'upsetting'.[2]

*In 2000 a public enquiry was held into events at the Bristol Royal Infirmary cardiac surgical unit where two senior, experienced, cardiothoracic surgeons continued to perform the 'switch' operation on babies despite their audit demonstrating very poor outcomes. In the same year it was revealed that pathologists at the Alder Hey Hospital in Liverpool had been removing whole organs from babies at autopsies and storing them for possible future examination. The Bristol parents were not fully informed about the poor results from the unit and the Liverpool parents were not asked permission for the removal and storing of organs.

9. *When obtaining informed consent, the doctor should be particularly cautious if the subject is in a dependent relationship to him or may consent under duress.*
10. *In case of legal incompetence (including minors), informed consent should be obtained from the legal guardian in accordance with national legislation.*

In addition to the above guidelines based on the Declaration of Helsinki, there are a number of conditions that apply purely to training, which would seem to maximise ethical practice:

- All reasonable measures should be taken to ensure each training exercise using patients is as safe as possible.
  These measures may include appropriate monitoring, supervision and the use of alternative training methods to ensure a trainee is fully 'prepared' to perform a technique on a patient (theoretical teaching, background reading and familiarity with the principles and complications of a technique, use of models and simulators, learning each single component of a complex task etc.).
- Trainers have a duty to maintain the skills they teach, and to stay abreast of changes and advances in the subject they teach.
- Any trainee who is ultimately found to be unsuited to anaesthesia should be persuaded to leave the specialty immediately.

Benefits to patients from participating in the training of junior anaesthetists

There are several ways in which the ethical similarities between participating in non-clinical research and in training differ: the most important is that there is no obvious benefit to the individual research subject (except sometimes pecuniary), but being involved in training can provide significant benefits to a patient:

- The patient has the advantage of there being a 'second pair of hands' during his or her anaesthetic.
- Each trainee has skills and experience which will be different from the trainer, so the patient benefits from their pooled knowledge, and often the trainee's general medical and surgical experience will be more recent than that of the trainer.
- The patient has the benefit of a pre-operative visit from two anaesthetists, so can hear and consider the information conveyed on two occasions. Thus there are two separate opportunities for questions, and time for the patient to digest and consider the information relayed between these visits. In addition, many people find 'an eminent consultant' too intimidating to discuss their concerns with, but may find a junior anaesthetist more approachable. (Obviously this benefit depends on juniors seeing patients from their training lists beforehand, and being familiar with what is involved in the procedure the patient is to undergo.)
  Many patients appreciate being able to help with medical education, and the patient who is finding the 'hospital experience' intimidating and disempowering

> *"It was a teaching hospital, so you felt you were giving something back when students poked and prodded and asked interesting questions"*
> R. Milnes, The Times, February 8th 2001

may find that the opportunity to 'help train doctors' makes that person feel useful 'within the system', and it can help to restore a patient's confidence in her dealings with health professionals. Most people derive some satisfaction from the knowledge that they have helped in training the 'next generation of doctors', since these will be the people caring for them, their children and grandchildren in the future.

- Many consultants who regularly teach juniors find that this teaching improves their own practice and makes them refresh their knowledge regularly. It is obviously an advantage to the patient if his consultant anaesthetist benefits from his involvement in training.
- A junior in training usually provides initial out-of-hours care. It is of benefit to patients if the junior called to see them at night is familiar with their history.

Optimising ethical practice: responsibilities of trainees

The ethical and legal burdens implicit in learning procedures on patients must be shared between the trainee and the trainer. Junior anaesthetists are doctors, with a duty of care to their patients, and to secure the well-being of their patients during a training exercise the usual principles of 'good practice' should be adhered to. In addition, the following suggestions may be helpful:

- Trainees must understand the theoretical knowledge underlying a procedure, before attempting to learn it 'in real life'.
- Trainees should not attempt a new procedure unless they are happy about the methods, the level of supervision and are aware of potential complications.
- Trainees should have a low threshold for asking their trainer to take over if they become unhappy about a new procedure whilst carrying it out.
- Many of the benefits to patients of being involved in training depend on their meeting the trainee beforehand during a pre-operative visit, and the trainee being present throughout their whole anaesthetic.
- Trainees should balance the risks to the patient of their learning the procedure against the value they will derive from the training exercise.

It should be noted that, under UK law, an equal standard of care is demanded from a senior anaesthetist and a new junior anaesthetist. Inexperience is not a justification if problems arise.[3]

Most of the above issues apply to all medical training, but in anaesthesia, training is often on an unconscious patient, and there is the potential for disaster from even the briefest anaesthetic, for the smallest operation in a previously fit patient. Thus it could be argued that anaesthetists must have a much lower threshold for abandoning a training procedure than general medical specialists, and must be more aware of their patient's views, since the patient may not be in a position to withdraw consent once a procedure is underway.

Currently health professionals can decide how much information to give their patients, provided that they inform them of all 'significant' risks. At present, if their practice is in keeping with that of a body of their colleagues, then they can use their clinical judgement to decide how to approach training juniors on patients. However, there is a general move towards more choice and transparency in the NHS, driven by the media, the government, patients' organisations and a number of high profile medico-legal cases. In the future the legislation which applies to medical students performing procedures on patients may well be adapted for junior doctors too, and in order to prevent the imposition of impractically stringent limitations, anaesthetists need to address the ethical and legal issues surrounding training 'the next generation'.

The patient will have redress in a court of law if they suffer injury for which their anaesthetist is responsible. They are entitled to an absolute standard of care irrespective of who performs the procedure. Consultants and trainees must always bear in mind that inexperience is not sufficient excuse for failing to reach that standard.

Trainers must find ways to teach and supervise that protect their patients from the under-performance of trainees during the early stages of learning new procedures. There has been little research to quantify these risks. At present the question of how to manage the consent for teaching remains an issue of professional judgement in each individual case.

References

1. World Medical Association. *Declaration of Helsinki 18th World Medical Assembly*. Helsinki, 1964 (latest revalidation 41st WMA, Hong Kong, 1989).
2. Brattebo G, Wisborg T, Solheim K, Oyen N. Public opinion about different approaches to teaching intubation techniques. *British Medical Journal*. 307:1256–1257.
3. Wilsher v Essex Area Health Authority. 2 WLR, 425, 3 BMLR 37, C.A., 1987.

Further reading

Brazier M. *Medicine, patients and the law*. 2nd ed. London: Penguin Books, 1992.

McHale J, Fox M, Murphy J. *Health care law: Text cases and materials*. London: Sweet and Maxwell, 1997.

6

The non-technical skills of anaesthetists

Georgina Fletcher and Ronnie Glavin

Education is about teaching people how to behave

> *Education does not mean teaching people to know what they do not know. It means teaching them to behave as they do not behave.*
>
> Ruskin

We are in the business of transforming people. The end product of that transformation should be someone who behaves as an anaesthetist. The first step towards this is to identify the components that go to make up that behaviour. This then allows both trainers and trainees to be clear about what they are trying to achieve through the education process. So we must try to identify what it is that makes someone a good anaesthetist; how does that person behave? Then we have to break this down into the different areas of knowledge, skill and attitudes that make up this behaviour.

If you think about the universally respected, good anaesthetist you will probably conjure up a picture of someone who is technically adept and medically up to date. Importantly, you will also probably be thinking about someone who always seems to be aware of how their patient is doing. They anticipate problems before they occur, they can manage a crisis smoothly and safely without panic, they work well with all the other people in theatre and generally get the best out of their team. At the end of the day, it is this behaviour that you want to help trainees to develop and, interestingly, only the first couple of items relate to theoretical knowledge and practical skills, the rest relate to what can be described as non-technical behaviour.

The day to day practice of clinical teachers pays considerable attention to the theoretical knowledge that underpins anaesthetic practice and the practical skills and procedures that are required in the routine management of patient care. These can be thought of as technical skills. Yet, research shows from analysis of critical incident reports, observations from operating theatres and simulation training[1] that while necessary, these technical components are not in themselves sufficient to ensure safe performance. To be good anaesthetists, trainees need additional skills and abilities, to translate theoretical knowledge and practical skills into effective patient management, e.g. team working and decision making. It is these more generic behavioural competencies that we call non-technical skills.[2,3]

While explicit discussion of non-technical skills might be relatively unusual in anaesthesia, it is common in other industries such as aviation, where they have long been recognised as being important for safe operations.[4] Indeed, the term non-technical skills has been borrowed from the European aviation training community, which requires that non-technical skills training is given to flight crew on an initial and recurrent basis.[5] This ensures that skills are maintained, associated safety related attitudes are reinforced and developments information about the effect of new equipment or from incident analysis, is shared to keep everyone up to date.

Recognising that there is more to being a good anaesthetist than just high technical proficiency is the first important step to beginning to teach this to our trainees. The next step is to accept that there are components of this 'other' aspect, the non-technical skills that can be addressed explicitly through training. As we have already said this has been successfully demonstrated in aviation where training of non-technical skills under the banner of Crew Resource Management training has been shown to have a positive effect on safety.[6] This means it should also be possible for us to address them explicitly in anaesthesia, and indeed an increasing number of institutions worldwide are doing this through simulation-based training (see Chapter 21).

What are the non-technical skills that are important in anaesthesia?

While we have said that non-technical skills are fairly generic, and even similar across professions, when trying to identify which skills to train it is important to realise there can also be differences. So, although non-technical skills overlap from one industry to another at a broad category level, the components of these categories will be specific for each industry and so require a rigorous process of identification.[7] Essentially the process behind this is no different from trying to identify the technical skills necessary for a task. We just need to adapt the techniques we use to allow us to isolate the non-technical competencies rather than technical ones. There are now a number of international research efforts investigating non-technical performance in anaesthesia.[8,9] And so there should be increasing amounts of material available for you to use in developing your teaching curriculum. In our work we use a technique known as cognitive task analysis to identify the non-technical skills of anaesthetists and from the results of this have developed a taxonomy of anaesthetists' non-technical skills.[1,2] This skills taxonomy and its associated behavioural markers is

The non-technical skills of anaesthetists:[2]

- Situation awareness:
  - Collection of information
  - Assimilation of information
  - Projection of what will happen
- Decision making
- Task management
- Team working

currently being evaluated. The four main skill categories we identified are shown in the box. These are then broken down into lower level skill elements to understand each area in more detail. For the purposes of the next section we will focus on these four categories, but in your training course would probably want to be more specific. Other things you could think about including might be leadership, communication skills and stress management.

How should we teach these skills and behaviours?

We tend to use the term non-technical skills as a generic term to cover all aspects of non-technical performance. However, as with technical skills, we actually need to consider the three areas of competence: knowledge, skills and attitudes. So, as trainers we need to know: what are the skills anaesthetists need to develop to behave in a safe and effective manner? What knowledge is needed, so the trainees know why these skills are important and what factors effect them, and what attitudes should we be endorsing to support safety? We will address each of these areas in turn.

What non-technical skills related knowledge should we teach anaesthetists?

Learners must know from an early stage of training that:

- They must practice safely.
- They must be effective.
- They must adopt a professional approach to work.

The first step is to understand the major concepts. We want our learners to behave in a safe and effective manner. We want them to adopt a professional approach to their work. This must be made explicit from an early stage in the novice's anaesthetic career. During the introductory sessions we can not only say what we want learners to do technically, but can say why we want them to behave in that way. We can stress the importance of these concepts in the effective delivery of patient care. Many anaesthetic departments will provide a list of topics that have to be covered. Most of these will concentrate on technical skills — the knowledge and practical skills specific to anaesthesia. While it is completely appropriate for novices to concentrate on learning these skills, we must not forget to include the concepts associated with non-technical skills.

Make explicit how learners should behave by covering these issues in the introductory programme:

- Safety
- Standards
- Professional approach

Also include:

- Team working
- Team roles
- Individual responsibilities within a team

An important question is when should formal teaching of things non-technical begin? In Scotland we introduce formal teaching in the third year of the UK seven-year training programme. We chose this time because we felt that trainee anaesthetists need to acquire a sufficient grounding in both technical skills and clinical experience in order to assimilate non-technical skills training effectively. Our course consists of a series of short presentations where we introduce the following concepts:

- Factoring the human into safety — a look at the different types of error that can arise and the circumstances in which they arise.
- Situation awareness — knowing what is going on around you, finding the most appropriate information and building effective mental models from that to allow effective decision making.
- Leader and follower roles in the team — when and how to lead and when and how to be an effective follower.

Following each short presentation (around 15–20 minutes) we want the learners to consolidate the concepts by making use of them. We do this by giving them small group exercises to carry out. Some of these will use material that we provide but we also encourage our learners to use challenging clinical episodes, near misses or critical incidents of their own during these sessions. Case based discussion is the most effective way to teach this material.

What does the anaesthetist need to know in order to develop non-technical skills?

- What are the responsibilities of the anaesthetist as a member of a team?
- What is the anaesthetist expected to bring to a team in the way of knowledge and skills?
- What are the roles and responsibilities of the other members of the team?

How will the novice anaesthetist identify those who are to provide such skills?

When anaesthetists work they use both technical and non-technical skills. Part of the concept that we are teaching is the relationship between them, including the limitations of each set of skills, and how the two aspects of performance can influence each other. For example, there is no point in being a great team worker if you do not have the technical skills to fulfil your role in the team, but conversely, without good team working skills it is very difficult for the purely technical expert to function effectively, especially in a crisis.

Technical skills alone are insufficient to ensure good quality patient care but non-technical skills themselves are no substitute for a lack of technical expertise.

Although we introduce *formal* teaching of non-technical skills (in the sense of an organised course) after a minimum of two years clinical anaesthesia, this does not imply that the concepts are not introduced before then. The advantages of a formal course, with activities to support and consolidate the concepts, are that an overall sense of structure can be introduced. However, during clinical work (or simulated clinical work) situations will arise where the concepts can be pointed out to the learners. For example, if the learner asks the surgeon for information on the state of the surgical field, the concept of obtaining information and building up mental models, and the importance of this concept, can be discussed. Opportunistic teaching, therefore, can and should supplement the formal courses. Indeed, as trainers we should equally be aware of trainees learning from our non-technical behaviour as much as our technical behaviour. Therefore, it is very important that standards of good practice with all their safety related aspects are demonstrated by everyone who comes into contact with the trainees. By drawing attention to the safety (and other) benefits of their behaviour trainees will start to understand why they are important. Discussion of cases can also help increase a trainee's knowledge about what skills and strategies for applying them are useful in which situations.

What non-technical skills should we teach anaesthetists?

Understanding the concepts is important but not sufficient. Good behaviour relies on these concepts being applied appropriately. So trainees not only have to know how they should behave, they actually have to behave that way. For example, in the technical area it is not enough for a trainee to know about putting in an epidural, they actually have to be able to manage the patient this way. Similarly, in the non-technical area, it is not enough to know about the different stages of making a decision, they actually have to be able to make decisions in practice. For this to happen, trainees have to develop the non-technical skills of the type listed earlier. During our formal course we make use of our high fidelity simulator to give our trainees the opportunity to apply the concepts they have been taught in the small group sessions, and so practice their skills. However, simulation is only one way to give trainees opportunities to try out these skills. Other recognised techniques include role-playing scenarios, problem solving games, team games, etc. Generally, it is in the clinical area that most learners will develop these skills.

The role of the teacher in these circumstances will differ from that for 'Knowledge'. In that role the teacher is seen as someone who imparts knowledge, whereas, in theatre the senior anaesthetist has to act more in a coaching role — they become a trainer. This role becomes easier to fulfil if the trainer and the learner have a common vocabulary that allows them to review events and the performance of the trainee.

Let us use situation awareness as an example of a non-technical skill. Situation awareness is the ability to know what is going on, and the ability to use this mental model to project what may happen. This requires the individual to gather information, assemble that information into a coherent pattern and to use that pattern to calculate what may happen. One way this skill can be developed is by asking trainees to state what they think is happening, and then asking them to justify why they have come to that conclusion. During routine maintenance of anaesthesia when everything is stable the learner can be asked 'What do you think could be going on if the heart rate were 120 beats per minute just now instead of 90?' For each condition listed, the trainee could be asked to list what the patient would look like, what the blood pressure would show, what the end tidal CO_2 would show, etc. As learners become more experienced they can be asked to think ahead about what could happen and, to show the links to decision making, to state what actions they would take if different situation arose. (See also Chapter 11.)

This kind of coaching can take place not only during events but can also take place after events, especially critical incidents. For example, following intubation of a patient with an unpredicted difficult airway the learner could be asked to review his or her actions. This would include technical skills — the protocol for managing the difficult airway, knowing what airway adjuncts should be available, being able to use these adjuncts and confirm successful placement of the endotracheal tube. Non-technical skills should also be addressed — informing the assistant of the difficulty, asking for the appropriate equipment to be made available, changing plans in view of the changing circumstances etc. If the learner has to be 'rescued' by the teacher then both the technical and non-technical skills should be addressed after the case.

Perhaps an important point to make at this stage is that, as with technical performance, not all trainees will have the same natural aptitude for non-technical skills and individuals may develop at different rates. Some trainees may be better in the social areas like team working, but less organised in their task management. Others may be really good in the cognitive areas, like decision making, but be very uncomfortable discussing a case with the surgeon. In these situations it is important to be able to recognise where a trainee's strengths and weaknesses lie, so that you can focus the necessary resources on the critical areas.

These are just some examples of the ways to help trainees develop their non-technical skills. There are numerous resources available that can be used for guidance on this sort of skill development. For example, there is a large volume of literature available on team working and team training. While not all written specifically for anaesthesia, as said before, many of the concepts are generic and the training methods employed could certainly be adapted for use in anaesthesia.

What non-technical skill related attitudes should we encourage in anaesthetists?

We can describe attitudes as a propensity to act in a certain way. Our attitudes towards something that will strongly influence whether or not we adopt related behaviours. Thus with respect to non-technical skills, associated attitudes we would hope to see

Some general principles of good non-technical skills training:

- It should be based of a sound theoretical background but in the context of the technical task.
- It should make use of examples and scenarios directly related to anaesthesia.
- Trainers should themselves be familiar with the material and concepts, and should be prepared to 'walk the walk'.
- Safety should be supported through the behaviours and attitudes of the whole department.
- It should be evaluated to ensure it is effective and should be kept current with the latest issues and challenges.

would value team working, encourage openness about errors and generally support activities associated with safe practice such as vigilant monitoring, good planning and following professional standards. Indeed, it is possible to see that one of the main aims of a non-technical skills training programme is to encourage trainees to adopt these values, because development of good safety attitudes is very important if trainees are to use the skills they have been taught.

We have previously addressed the non-technical skills we want trainees to develop and why these skills are so important, this section is about encouraging learners to use these skills. For this we move even further from the classroom.

Habits

One way in which we can influence behaviour is to encourage the trainee to develop good habits. By this we mean to carry out certain routines in such a way that they become automatic, e.g. checking the label of a drug ampoule before drawing up a drug, checking the label on a syringe before administering a drug. Other examples would include: checking the anaesthetic machine, preparing the routine drugs and equipment etc. Once learners have been taught these routines they are likely to become habits if their performance is encouraged, so the behaviour must be reinforced by repetition.

Values

We have touched upon concepts such as safety and the professional ethos of commitment to high standards of patient care. Introduction of these must be followed up by an acceptance of these concepts, and by behaving in a way that is consistent with them. The two recognised techniques for bringing about adoption of such values are *reinforcement* and *human modelling*.

Reinforcement

In the sections on knowledge and skills we have described the importance of making these aspects explicit and teaching the learners how to act. However, unless the learners see other trainees or consultants act in a consistent manner, then these behaviours are less likely to be adopted. If we wish to promote the habit of checking drug ampoules then we, as teachers, must do this ourselves.

Human modelling

Consistent performance by the consultant body is more likely to result in values being adopted. However, the actions and behaviour of some key individuals will have a particularly strong effect. Certain individuals will be taken as role models, and the behaviour of these individuals is more likely to be taken up than the behaviour of others. In the case of trainees the role models may not be consultants, but may be more senior trainees.

Feedback

There are two aspects of feedback that are particularly important in helping to develop the non-technical skills associated with good anaesthetic practice.

1. Allowing, and indeed encouraging, learners to follow up the consequences of their actions. For example, decision making is only going to improve if the learners first of all find out what the results of their decisions were.
2. Encouraging the learner to review his or her performance. If learners are to develop their behaviour, then we need to have a system that allows us to identify the components of behaviour. For example, it is of little help to a learner to say "You need to make better decisions". How does a learner do that? If, however, we can help a learner review his or her performance and come to realise that not enough options were being generated, then have helped identify why decision making was poor. The answer to this is to develop a system of behavioural markers that will allow us to carry out such a role and so help learners reflect more accurately upon their performance and so consequently make improvements in their behaviour.

The most important thing to remember is that non-technical skills training it is about teaching people to behave in a safe and effective manner at all times, in all situations and to be good anaesthetist; so its goal, which is to enhance safety, should not be forgotten.

References

1. Fletcher G, McGeorge R, Flin R, Glavin R, Maran N. The role of non-technical skills in anaesthesia: a review of current literature. *British Journal of Anaesthesia*. 2002;88: 418–429.
2. Fletcher G, Flin R, McGeorge R, Glavin R, Maran N, Patey R. *Final Report: Development of a behavioural marker system for Anaesthetists' Non-Technical Skills (ANTS).* Aberdeen UK: University Of Aberdeen Grant Report, 2001.
3. Fletcher G, Flin R, McGeorge R, Glavin R, Maran N, Patey R. *Development of a prototype behavioural marker system for anaesthetists' non-technical skills*. In review.
4. Helmreich RL. On error management: lessons from aviation. *British Medical Journal*. 2000;320:781–785.
5. Flin R, Martin L. Behavioural markers for CRM: A survey of current practice. *International Journal of Aviation Psychology*. 2001;11(1):95–118.
6. Helmreich RL, Merritt A. *Culture at work in aviation and medicine.* Aldershot: Ashgate, 1998.
7. Glavin R, Maran N. Development and use of scoring systems for assessment of clinical competence (Editorial). *British Journal of Anaesthesia*. 2002;88:329–330.
8. Gaba D. Human error in dynamic medical domains. In: M. Bognor, editor. *Human error in medicine.* Mahwah, NJ: Lawrence Erlbaum Associates, 1994;197–225.
9. Helmreich RL, Schaefer H-G. Team performance in the operating room. In M. Bognor, editor. *Human error in medicine*. Mahwah, NJ: Lawrence Erlbaum Associates, 1994; 225–253.

7

Problem-based learning (PBL)

Philip Liu and Letty M. P. Liu

There is no consensus as to what constitutes Problem Based Learning (PBL) but the important distinction between PBL and the lecture format, is that it involves active learning by enquiry.[1] Customarily, a group of learners are set a task that requires them to divide tasks between them and then work as a group to develop solutions. The group are assisted by a facilitator, who may not know anything about the topic being researched. The role of the facilitator is to help the group co-operate in the work and progress to a conclusion.

The key elements of problem-based learning:

- Problem oriented
- Active learning by enquiry
- Small groups (less than 10 learners and one facilitator)
- Learner centred
- Case-based — clinically relevant

Learners work together to achieve an understanding of the issues presented in the case, and are required to come to the session ready to engage in a well-researched discussion. This active role is critical to the PBL process in contrast to lectures, where the learner is the passive recipient of information selected by the lecturer. Patel and her colleagues compared students undertaking a conventional curriculum at McGill with those undertaking a PBL curriculum at McMaster.[2,3] They found that the PBL students used a system of reasoning that had been explicitly taught. They were surprised to find similar patterns of reasoning emerging spontaneously in conventionally taught students, who had not been instructed in how to reason. This is important to those who want to use PBL for teaching medical postgraduates from different backgrounds, as it suggests there will be no fundamental difference of approach between those with previous experience of PBL and those without. PBL also differs from the traditional format because it mandates using small groups of participants, rather than a large audience in a lecture hall. Working in small groups brings into play a number of instructional issues related to small group dynamics. Small groups create pressure to perform satisfactorily in front of peers. The roles that group members assume are important. Who controls the agenda? Who are the information providers? Who challenges whom? Who is the timekeeper? Do all the members of the group participate? PBL shares these characteristics with a variety of learning exercises that can be used in small groups.

Problem-based learning (PBL) has been introduced in many medical schools as part of an effort to improve learning.[4–7] It is less used in postgraduate specialty education. It has been advocated as a way of developing the student's problem solving in clinical practice. Some schemes have involved early, close integration of basic science and clinical practice. Enthusiasts believe that the use of scientific reasoning in clinical situations will be encouraged, and that this will also result in improved problem solving. Some evidence, however, suggests that science based and clinically based reasoning use qualitatively different approaches.

The evaluation of students in a PBL program also differs from that of students who are taught through lectures. In PBL a conventional assessment against outcomes that relate principally to knowledge and understanding, as demonstrated in the discussion, can be made. PBL also permits the faculty members to make an assessment of each member of the group based on the individual's contribution to the discussion and performance in the group.

The PBL format for clinical medicine is based on a case involving a patient or a situation that the learner may encounter clinically.

Using cases as the centrepiece of PBL ensures that the issues and content of the educational experience will be clinically relevant. The case should be presented in a way that leads learners to the key topics and dilemmas. The most engrossing cases are based on real life medical conundrums. Forcing students and practitioners to confront case-based problems provides the learners with a model that they can adapt to their future medical practice. Thus, PBL is a model that encourages lifelong learning. It is a format that simulates the problems, realities and nuances of medical practice.

A distinctive characteristic of human nature is that we are intrigued by problems. PBL capitalizes on this innate fascination. We are driven to attempt to arrive at a solution, and we are exceptionally pleased when we feel that we have succeeded in finding one. In the PBL format learners can pursue topics to the limit of their enthusiasm. The choice of resources is the prerogative of each student. The learner is responsible for preparing for the discussion session, so that they can answer questions and defend strategies for solutions that may arise. This exposes them to the task of managing and using information and educational resources. With the ubiquitous presence of computers in this information age, PBL beckons the learner to become a master of modern informatics. PBL participants may use textbooks, consult with colleagues, use computer-aided literature search programs, journal articles, or other educational resources. The learner/discussant ultimately should be prepared to discuss the issues in a collegial setting with peers. The PBL discussion session fosters teamwork in problem solving and the concept that learners can learn from their peers.

Where does PBL fit into the scheme of things in clinical postgraduate education?

Problem Based Learning:

- Uses the learning skills appropriate to small groups.
- Conforms with good teaching practice for adult learners.
- Exercises skills in information retrieval and management.
- Requires learners to discuss and debate with one another.
- Concentrates minds upon researching and resolving a target task.
- Demonstrates reasoning skills.

As foreign a concept as this may be, PBL can be fun! Ideally this should be the goal of every PBL session.

The objective of PBL is to permit the learners to attain insights by the process of their own deliberations, and thus permit them to experience the thrill of discovery learning. This active engagement in the intellectual process is a valuable asset. In cognitive psychology the distinction between the novice and an expert is that the expert can categorize new knowledge and material. They know what they know, they know what they do not know and they can recognize what is new in relation to the old material. They can evaluate what the risk/benefits ratios are of a number of practice options, and will make medical judgements and decisions on the basis of weighing alternative strategies. Not only must the learner read and absorb new information, but they should also learn to make judgements and decisions based on the case data and the newly acquired data in order to solve the problems arising from the case. The process is much more complex than merely reading information or listening to orally presented material. It is much more aligned to the reality of medical decision making in practice.

Undergraduates are not 'in the loop' for patient treatment, and a preoccupation of educators is to supply them with realistic opportunities involving or simulating clinical practice. In undergraduate medical education the learner's decision making is always somewhat artificial in that the actions are not played out. The patient is not injured by a bad decision. Postgraduate specialist trainees encounter many opportunities to make real decisions.

Does PBL have a place in postgraduate education?

PBL is useful both in letting trainees explore their decision-making, and as a way for the supervisor to assess the processes by which they develop an opinion. It also establishes a 'community of cooperative learning' within the department.

The role of the faculty member

Curriculum development for problem-based learning is the same process as that for developing traditional curriculum. Content areas are designated and exercises developed to address those areas of content. The strength of case based learning is the immediacy of the clinical relevance, because the content is framed in the context of a clinical case. Even though the core content may be basic science topics, they have been successfully transmitted into a case based curriculum.

The faculty members' role is that of a moderator/facilitator who assists the learner/discussants in identifying the key issues of the case and helps the group by facilitating and moderating the discussion and monitoring the quality of the discussion. They are there to facilitate the learners in a process of self-discovery. The moderator/facilitator may also make an assessment of the trainees' performance in their role in the PBL group.

Moderation and management of small groups are skills that the discussion leader must master in order to be effective. They are distinctly different from lecturing skills. Adult learners are much more comfortable and efficient in their learning when they can understand and control the learning process. The faculty member serving as the moderator/facilitator must be prepared to allow the learners to control the agenda. The learners can set their own pace and dwell on the issues that they choose. The moderator/facilitator must allow the learners to identify the key issues in the case. The faculty member relinquishes control to the learners. In particular, faculty must overcome the urge to be information providers and give mini lectures. Trust the learners; they will come through for you! This adage should be the motto for PBL faculty.

Pros, cons and controversy in problem-based learning

Many studies have compared medical school graduates trained in the traditional educational programs versus programs that emphasize PBL.[5,7–9] Almost all show

> Many of the arguments presented in support of PBL originate from the belief that PBL produces better problem solvers. Unfortunately, most of the papers on this topic have not been able to confirm this thesis. Supporters also contend that active learning in a clinical context makes medical material easier to recall.

little difference in the outcome of graduates, when the results of licensing examinations and similar evaluation instruments are used. A few of these studies indicate that graduates from traditional programs perform slightly better. The happiness index amongst students in PBL programs has been shown to be higher than that of students in the traditional curriculum.[9,10] This is a major asset of PBL. A review of outcomes indicated that medical students at PBL schools enjoyed their experience more. In medical schools where two tracks exist, the PBL track is more popular. This attitude towards the learning environment is an important factor, and should not be under-estimated in its promotion of a positive educational result.

Surveys of graduates of medical schools that used the PBL format indicate that they do not feel disadvantaged.[9] Other studies suggest that PBL graduates are more sensitive to patients' psychosocial needs. Studies examining ratings of residents by residency directors indicated that PBL graduates were rated higher.[9] Many studies claim that the faculty preference is for PBL, because the format provides greater and richer interactions with students.

"Well, another day — another learning experience."

One purported advantage of PBL is that student retention of material is better because they learn material in a clinical context.[11] Although one article questions the 'backward reasoning' nature of PBL learners (emanating from hypothesis to data), as contrasted with more efficient process used by experts.[12] The author does concede that there are suggestions that retention of information in the context of patient problems is higher. This degree of recall is one of the primary virtues of PBL. PBL students generally attempt to develop an understanding of concepts rather than memorize lists.[10, 12] This should create a more able clinician.[13] Clearly, PBL students use library resources more and employ a wider variety of educational resources.[5, 7, 9, 10]

Albanase and Mitchell have conducted a meta analysis of over a hundred studies of PBL published between 1972 and 1992.[9] They conclude that:

"Compared to conventional instruction, PBL, as suggested by the findings is more nurturing and enjoyable: PBL graduates perform as well, and sometimes better, on clinical examinations and faculty evaluations; and they are more likely to enter family medicine. Further, faculty tend to enjoy teaching using PBL. However, PBL students in a few instances scored lower on basic sciences examinations and viewed themselves as less well prepared in the basic sciences than were their conventionally trained counterparts. PBL graduates tended to engage in backward reasoning rather than the forward reasoning experts engage in, and there appeared to be gaps in their cognitive knowledge base that could affect practice outcomes."

It is safe to conclude that there is no clear-cut answer in support or against PBL as opposed to traditional teaching formats. The data suggest that, though there are small differences between the formats, they are not nearly as great in quality and or quantity to justify the massive educational revolution towards PBL.

A problem for the hospital, service based, postgraduate teacher is that PBL is very time consuming to prepare. Many PBL routines are extended over a number of sessions because an important part of the exercise is finding and sorting through literature to develop an opinion. In the workplace it often turns out that members of the group are unable to attend all the sessions. This challenges the department to allow trainees regular 'protected' teaching time. In postgraduate and continuing education doctors are likely to attend sessions that they enjoy.

The American Society of Anesthesiologists (ASA) has used PBL for anaesthetists continuing education since 1992 and has made innovative modifications to classic PBL in order to make it more useable in a time constrained postgraduate setting. The learner/discussant is given a list of references in a sealed envelope. By enclosing pertinent references for the case, the learner/discussant can save time by opening the envelope. Alternatively, they may choose to study independently and seek solutions to the case problems by using their own educational resources. Prior to the discussion session, the learner/discussant may then choose to compare his references with those provided by the moderator facilitator. An extraordinary statistic is that >98% of ASA problem based learning discussion (PBLD) participants indicated on evaluation forms that they plan to take another PBL session and would recommend PBL to a friend (authors unpublished data).

Like most innovations, PBL will have to stand the test of time. However, we can certainly use it in our curriculum and have fun with it. In fact, anaesthesia is a pioneer among medical specialties in its use of PBL and the specialty contributing in an active manner to the continuing dialogue.

Implementing a problem-based learning program

Faculty development

The key to the success of the ASA PBLD program* was the inculcating of the philosophy of PBL into the faculty and the faculty development program that was undertaken. At the annual meeting prior to the ASA PBLD program, faculty development workshops are held. There are discussions on the philosophy of PBL and practice PBL sessions are conducted where faculty can practice their moderating/ facilitating skills and receive constructive feedback on their performance. A PBL guide has been developed and is provided to the faculty. This outlines the critical moderator/facilitator skills that we encourage faculty to develop. A format for a faculty development meeting is given in Appendix 1 and an example of a scenario for PBL is shown in Appendix 2.

A programme of faculty development is the key to successfully introducing a programme of problem based learning discussions.

References

1. Walton HJ, Matthews MB. Essentials of problem-based learning. *Medical Education*. 1989;23:542–558.
2. Patel VL, Evans DA, Kaufman DR. Reasoning strategies and the use of bio-medical knowledge by medical students. *Medical Education*. 1990;25:527–535.
3. Patel VL, Groen GJ, Norman GR. Effects of conventional and problem-based medical curricula on problem solving. *Academic Medicine*. 1991;66:390–389.
4. Tosteson DC. New pathways in general medical education. *New England Journal of Medicine* 1990;322:234–238.
5. Neufleld VR, Woodward CA, MacLeod SM. The MacMaster MD program: a case study of a renewal in medical education. *Academic Medicine*. 1989;64:423–432.

*ASA PBLD books have PBL cases, case objective, model discussion outline and references for each case presented at the ASA annual meeting. A case writing guide is also included. Books may be obtained from the American Society of Anesthesiologists, 520 N. Northwest Highway, Park Ridge, Illinois 60068-2573, USA.

6. Schmidt HG. Problem-based learning: rationale and description. *Journal of Medical Education*. 1983;17:11–16.
7. Kaufman A, Mennin S, Waterman R, et al. The New Mexico experiment: Educational innovation and institutional change. *Academic Medicine*. 1989;64:285–294.
8. Schmidt HG, Dauphinee DG, Patel VL. Comparing the effects of problem based and conventional curricula in an international sample. *Journal of Medical Education*. 1987;62:305–315.
9. Albanese MA, Mitchell S. Problem-based learning: a review of literature on its outcomes and implementation issues. *Academic Medicine*. 1993;68:52–81.
10. Moore GT, Block SD, Briggs SC, Mitchell R. The influence of the new pathway curriculum on Harvard medical students. *Academic Medicine*. 1994;69:983–989.
11. Claussen HFA, Bousbuisen HPA. Recall of medical information by medical students and doctors. *Medical Education*. 1985;19:61–67.
12. Norman GR. Problem-solving skills, solving problems and problem-based learning. *Medical Education*. 1988;22:279–286.
13. Mitchell R, Liu PL. A study of resident learning behavior. *Teaching and Learning in Medicine*. 1995;7:233–240.

8

Educational supervision and mentoring

Chandra Kumar and Chris Dodds

Three perceptoral functions for learners:

- Clinical Supervision — providing clinical support and oversight to a trainee to ensure that patient care is always satisfactory.
- Educational Supervision — helping the trainee identify suitable educational objectives, planning suitable experience for achieving those objectives and making assessments of whether the objectives have been satisfied.
- Mentoring — supporting the trainee as an experienced and trusted advisor.

There are three perceptoral functions that clinical teachers must establish for their learners. Clinical supervision; the provision of clinical support, has been discussed in Chapter 2. This Chapter is concerned with the two other relationships.

- Educational supervision is about planning clinical learning and organising its assessment.
- Mentoring is a one to one relationship established between the learner and a trusted, senior colleague to help them avoid difficulties and cope with any problems that may affect the progress of their learning.

Trainers must understand these processes and ensure that they are carried out properly.

The educational supervisor

The passage from novice to expert is long and complex and requires educational support. Trainees will need to work in many different hospitals and departments and the task of educational supervision will travel from one trainer to another as key stages in their development mature. Their learning in each stage must be planned, suitable experience must be organised, assessment of the outcome of learning must be made and a report of that assessment must be recorded. The ending of one stage frequently overlaps the start of the next and each supervisor must be able to trust his or her colleagues to have conscientiously dealt with their share of the learning and assessment. A degree of mutual respect for other colleagues involved in the training is imperative.

The role of educational supervisors varies but includes some or many of these:

- Role model
- Teacher
- Assessor
- Facility regulator
- Educational opportunist
- Contract holder
- Career advisor
- Personal advisor
- Proxy employer

"Yes, darling! Mummy has to keep her hands lovely in case she ever wants to go back to brain surgery."

The relationship between a trainee and their educational supervisor may be very short or may last for their full period of training. This depends on the nature of the training, its complexity and often on the stage of training being entered. Trainees do not usually choose their educational supervisors, they are allocated to them, especially in the early years of their training. The supervisors will have undergone some training in their role, and they will have the skills necessary to complete the tasks ahead of them. In some countries they are appointed to all novice doctors and have the responsibility of recommending their trainee for full recognition as a registered doctor. This task determines the structure of their role in that they have to provide educational and clinical opportunities in which the doctor can be observed to perform to a prescribed standard.

An educational supervisor should always be a good role model for the trainee. They must understand and be competent in the technicalities of their phase of training. They must know what relevant experience is available and be able to match the trainee's duties with suitable experience. In specialties that work in "firms" or teams the consultant can control the trainees work and direct them towards the necessary experience but in anaesthesia the clinical supervisor usually has less control of work allocation and this is probably the greatest barrier to efficient learning in the course of real work.

The skills required of an educational supervisor

As a teacher

Clinical teaching. Supervisors need to understand the processes of workplace-based learning and assessment. They should strive to be exemplary clinical teachers. They will also need to have received advice and training on presentation skills, lecturing, small group activities and so on.

Planning learning. The educational supervisor must review the learner's prior learning, including previous supervisor's reports and plan some suitable learning objectives for this phase of the trainee's progress. They will usually deal repeatedly with the same clinical area and trainees at similar stages of training. They will know what is appropriate and what is achievable. The planning process must take note of the trainee's own assessment of their needs and understanding of their current level of competence. It must be tailored to the trainees individual career planning. It must take note of what future training and experience will be available for each individual trainee.

Maintaining the educational environment. The educational supervisor must establish and maintain an emotional environment conducive to learning. The environment should be one of honesty, trust and fairness. The supervisor should actively support learners and prevent confusion and blame when things go wrong. Clear statements to the rest of the clinical team about the lines of responsibilities especially for patient safety will reduce the systematic errors that may occur.

The necessary qualities of an educational supervisor:

In connection with their clinical work:
- Master of the relevant curriculum
- Proficient clinical performance
- Knows the available, relevant learning opportunities

As a teacher:
- Knows what the trainee needs to learn
- Able to help a trainee to plan their learning
- Able to assess the trainees clinical performance

Personal qualities:
- Excellent clinical role model
- Clear communicator
- High ethical standards
- High professional standards

As an educational entrepreneur

Skilled in opportunistic teaching: Educational opportunities occur in the trainee's everyday work and during routine contact with the supervisor. These often arise by chance but occur in almost everything the trainee does. The positive use of everyday events enhances the respect between supervisor and trainee and encourages reflection.

Being organised: There are deadlines for learning and testing key skills. These cannot be left to chance. Liaison with others may be necessary (resuscitation training for example) and this has to be time-tabled into the training scheme. Later in training, broad clinical experience becomes important and collaboration with medical and para-medical colleagues may be the only way to provide that knowledge. Supervisors can only remain effective if they foster these parallel avenues of experience.

As an assessor

Success as an assessor depends on skills learnt in the supervision and observation of many trainees. Other departments or individuals whose sole job is to assess trainees in clearly defined aspects of training may do some assessments of the trainee. However, interpretation of these assessments depends on the experience of the supervisor. The ability of an individual supervisor to assess all components of training at an equal and reliable level is open to question and it is likely that there will be a bias due to the differing experiences and enthusiasms of individual supervisors. Insight (or guidance) into this bias is necessary and will dictate when support from other colleagues is necessary. The clinical skills that are at the core of independent specialised practice must be most closely inspected and assessed, a process that requires expertise in those areas from the supervisor. The educational attributes that confer the ability to maintain lifelong learning skills are as essential but are often acquired from other medical paths and are usually more limited in the number of supervisors with these skills.

As a regular supervisior

The day-to-day duties of clinical supervision are shown in the box.

The task of supervising learning has four main elements:

1. *Direct learning:* The supervisor should be concerned with the content and process of training. They should be concerned with the development of the trainee's understanding and personal or technical skills. The trainee must develop an ability to learn directly from the task performed. The trainee must be able to provide effective feedback to the supervisor on their performance.
2. *Support of learning:* The supervisor must ensure that the trainee does not attempt to perform tasks that are beyond their experience or knowledge. Clear statements of the lines of responsibility and limits on acceptable variations will reduce the stress placed on trainees, and often prevent the apparent apathy such overwhelming stress can cause. Supporting learning will also include assessment of performance.
3. *Recording progress:* The supervisor has a responsibility to fully record all aspects of the trainees performance both good and bad. This should happen so frequently that it becomes routine. The quality of performance should be discussed with the trainee. This process of feedback and recording is the only safe way of providing evidence of satisfactory or poor performance.
4. *Managing clinical experience:* The educational supervisor must reconcile the needs of training with the demands of service work. The trainee must be allowed to undertake the work that they need in order to meet their learning objectives. This must not be at the expense of patient care or of the learning needs of other trainees.

Tying it together — the educational contract

A written contract or agreement between supervisor and supervisee is very valuable because it enables both the supervisor and supervisee to understand what aspects of work are essential learning experiences and what are not. It allows formal recognition by both parties of the full range of learning objectives. For instance, tasks such as administration and managerial training are important elements of training in anaesthesia and should be included in the educational contract as of equal importance to clinical ones.

The mechanics of supervision

The preliminary interview: At the start of the supervision the supervisor and trainee must meet. They will review the trainee's previous progress and agree a plan for the new period of learning. This plan will define the necessary practical experience and the parties will decide how they will judge whether the trainee's performance in the new areas is satisfactory.

Continuing supervision: The supervisor should informally discuss progress with the trainee whenever possible. There should also be more formal meetings to discuss how things are going. These should be frequent enough for the trainer to be fully in the picture about the trainee's progress. The period of supervision may include relatively formal assessments with recorded outcomes. There may be a requirement for particular tasks to be seen or undertaken. There may be a requirement to log a specific number of procedures. There should be continual feedback about progress to the trainee and they should never be in any doubt how they stand.

The review of progress: At the end of supervision is the meeting between the supervisor and the trainee at which the educational contract is reviewed to decide whether the core skills, competencies and attitudes required for promotion or progression have been acquired. This formal review should hold no surprises for either party. If there is likely to be dissent, or where performance has been below the acceptable limits it is prudent to have more than one supervisor present and to record clearly the decisions taken and their reasons. In some training systems this is a formal requirement before employment is renewed — the Record of In Training Assessment (RITA) in the UK. There are three possible outcomes.

- Progress is satisfactory. Transfer to the next year or grade is approved.
- Progress is generally satisfactory but there are some areas of underperformance or where some necessary training has not occurred. Progression can be allowed, provided that these poor areas are re-addressed and the following review confirms overall progress.
- General performance is so poor that progression is not possible, and repeating the training is advised.

Repeatedly poor performance may lead to the trainee being discharged from training. The consequences of this are so serious that an appeal is likely and due processes have to be in place to allow this to occur. True independence of the appeal panel from the original review board is essential for the trainee's sake. As there are likely to be employment issues at stake clear and comprehensive documentation throughout the total period of supervision is vital. In Chapter 17 we shall discuss feedback in the context of assessment and appraisal.

Mentoring

A 'mentor' is an experienced and trusted advisor

The aim of mentoring is to help the individual in achieving their full career potential but a mentor is a personal as well as a professional advisor. A clinical supervisor or 'faculty advisor' may become a mentor if the relationship develops in that direction.

"A good mentor seeks to help a student optimise an educational experience, to assist the student's socialization into a disciplinary culture, and to help the student find suitable employment."[1]
"Mentors are advisors, people with career experience willing to share their knowledge; supporters, people who give emotional and moral encouragement; tutors, people who give specific feedback on ones performance; masters, in the sense of employers to whom one is apprenticed; sponsors, sources of information about and aid in obtaining opportunities; models, of identity, of the kind of person one should be . . . ."[2]

Have you been mentored?

Many doctors have already used a mentor — they just didn't realise what was happening. Almost everyone has someone who took an interest in his or her welfare and development. You may not have thought of that person as a mentor, but you know that it was an important relationship and a valuable influence. A mentor has to provide a role model for the trainee at a particular stage of their development, with the implicit understanding that as they develop that relationship will cease. The choice of a mentor has to be made by the trainee (mentee) because it involves many personal as well as professional matters and a degree of personality matching is essential. The limited number of skilled and willing mentors is a problem, and some will find they have to limit the number of mentees they can support.

Why bother to have one?

Mentors help their protégés to develop new insights into their practice and learning needs as well as fresh perspectives on their training or continuing professional development. Mentors may act as a catalyst in the development of new approaches to clinical practice and to consideration of further training. The constructive support and alternative views of an experienced guide may lead to career choices or paths that would never have been considered and to new opportunities for a satisfying professional life that otherwise would have remained dormant. However, they don't work all the time and it is possible to have a 'malignant mentor'.

Qualities of a mentor

Some good qualities for a mentor:

- Empathy
- Understanding
- Genuineness
- Concern
- Attentiveness
- Openness

Good mentors are:

- Good listeners
- Good observers
- Good problem solvers

Not all consultants are suitable to be mentors. The ideal personal qualities are shown in the box. These will vary between supervisors making them suitable for some trainees and not others, or at one time of training and not another. The qualities of a mentor vary with the needs of the trainee, but there are some aspects that are commonly found. The roles of career enhancement, personal skill development and 'useful contact' formation intermingle and make a single list of attributes impossible. There are some general features however.

Why should I be a mentor?

- To share your knowledge and experience.
- For the personal satisfaction of helping.
- To help people develop to their full career potential.
- To influence new anaesthetists in ways that you think are worthwhile.
- To attract good students to your hospital.
- To keep in touch professionally.

Experience: The mentor has to be more experienced than the mentee and is selected because it is perceived that they are likely to be able to pass on their wisdom and use their expertise to help achieve the career development aims of the mentee. Usually the mentor is more senior and experienced although sometimes a near peer may be selected. Whilst for some trainees the mentor should be technically competent, for others political or management skills are more important. Whatever needs are perceived by the mentee, the mentor should be able to communicate and model appropriate and effective attitudes.

Openness: The mentor should be open. This means being explicit about what one is doing, how one is doing it and why. Openness also means being open to comment and criticism about one's own professional actions. Debate, even if heated, is part of this process of personal review and development.

A friend: It should be a pleasant and positive experience to share time with a mentor. They must relate to the mentee at many levels. They should inspire trust, have good interpersonal skills, be willing to help, be honest, empathic towards the problems of trainee, be frank in discussing professional development, and be capable of keeping a secret.

A sage: Mentors need a wide range of assessment, appraisal and counselling skills. The mentor must be able to identify the strengths and weaknesses of the trainee without becoming patronising or appearing critical.

Functions of mentors:
- Giving advice, career guidance and professional development.
- Acting as sponsor and advocate of trainee.
- Providing information on learning and career opportunities.
- Arranging appropriate professional experience.
- Identifying goals and professional values.
- Stimulating learning.
- Being a good role model.

As a mentor you should:
- Devote time to the relationship.
- Develop the relationship by informal chats over coffee etc.
- Listen patiently.
- Share your own experiences of success and failure but ... .
- Don't talk about yourself too much.
- Encourage your mentee to be self confident.
- Look for opportunities to give praise.
- Give critical feedback when necessary.
- Be sensitive to the possibility that you are being more intimate with your mentee than they wish.
- Never abuse your position.
- Be very aware of gender, racial, ethnic and religious sensibilities.

Mentors help mentees to:
- Establish themselves quickly in the learning and social environments.
- Understand the organisation of the medical establishment.
- Understand appropriate behaviour in different situations.
- Understand different and conflicting ideas.
- Develop values and an ethical perspective.
- Overcome setbacks and obstacles.

- Acquire an open, flexible attitude to learning.
- Enjoy the challenges of change.

Who was a useful role model in your medical school?
- Who helped you uncover and use your hidden talent or ability?
- Who helped you resolve a difficult situation in your life?
- Who challenged you to acquire a new direction in your life?
- What was it about these people who helped you?

The process of mentoring

The relationship between mentor and subject has to be negotiated. It may be brief, or it may become a prolonged relationship, and clearly defined end-points should be agreed as early as possible. The agenda depends almost entirely on the mentee. The task of the mentor is to respond to the trainee. The responses should help the mentee to see clearly for themselves. The mentor may play 'devils advocate' in helping the mentee to see themselves and their actions as others see them. Mentors should not tell. They should not judge. All advice should be offered within a framework of alternatives. Where the mentee presents with a big problem that is an obstacle to training the mentor should direct them towards sources of help, rather than seeking to solve the problem themselves. To become involved in direct action on behalf of a mentee undermines the relationship.

Mentoring is about challenging as well as supporting. Constructive criticism can help a mentee to face the need for change. The mentor has wider or different knowledge, experience, and skills than the mentee and should use these during a mentoring session. Mentoring is about helping.

Maintaining confidentiality is essential, and mentor and mentee should agree the boundaries of the relationship. Mentoring takes time: how much and how often should be agreed between mentee and mentor, as should the location. The process will have varying time demands over the course of the relationship; some periods may require

To promote mutual respect:
- Listen carefully to what your mentee has to say.
- Take your mentee seriously and be positive about their ideas.
- Don't tell your mentee what to do.
- Be frank and direct without being censorious.
- Suggest other mentoring relationships if necessary.
- Take their hopes and fears seriously — don't belittle them.
- Take trouble to meet where your mentee feels comfortable — on their home ground.
- Be encouraging.

"Have you tried drink?"

only a meeting for an hour once a month or so, other periods will need weekly meetings to achieve progress.

The only certain event in a mentoring relationship is that it will end. This may happen when the mentee has reached a stage when he or she no longer feels the need for regular contact. It is important to consider how the relationship will end. Mentor and mentee may agree to meet socially or less often, or simply stop completely. If it has been successful there will be cause for celebration and a sense of loss, and both should be considered in the final discussion.

Mentors should look back and review the relationship, considering its value, the original goals, and whether they were achieved. The goals may have changed over time as new aspirations were discovered.

"Mentoring gives a real buzz and makes me feel unbelievably good that somebody can learn and develop with my help. It has enabled my influence to spread and thus assisted the change process in a way which is more powerful than any other process I know."

Common to all mentoring is that mentees come to view things in a new light. Mentoring is about change — both responding to change in the environment and promoting change in the mentee. The basis of change is a new vision of the possibilities.

References

1. National Academy of Sciences, National Academy of Engineering, Institute of Medicine. Advisor, teacher, role model, friend: On being a mentor to students in science and engineering. Washington DC: National Academy Press, 1997;1. (Available on-line at http://www.nap.edu/readingroom/books/mentor.)

2. Council of Graduate Schools. *A conversation about mentoring: Trends and models.* Washington DC: Council of Graduate Schools, 1995.

Further reading

Morton-Cooper A, Palmer A. *Mentoring, preceptorship and clinical supervision: A guide to professional roles in clinical practice.* Oxford: Blackwell Science, 2000.

9

Learning by maintaining a 'Portfolio'

Shashi Kant Gupta and David Greaves

Professional training must prepare the doctor for a lifetime of independent learning. To succeed in this requires specific habits of mind. Important amongst these are the abilities to perceive where there is a need for learning, to plan that learning and to reflect upon the experience. This complex process allows the doctor to acquire new knowledge, skills and behaviours and to incorporate them meaningfully into their clinical repertoire.

Conventional postgraduate medical education does not specifically cultivate an approach to life-long learning. The proper use of a portfolio of learning may help do so. The curriculum for Anaesthetic trainees includes clinical and non-clinical, knowledge, skills and professional behaviours. Teaching, learning and assessment concentrates heavily on skill and knowledge. Anaesthetic teachers need a way to direct their trainees towards non-clinical skills and professional behaviour.

Logbooks

Benefits to the learner of keeping a log of experience:

- It helps shift the balance of power over learning from the trainer towards the trainee.
- It gives the learner, more control over the content and the approach to learning.

Repeated experience is needed to achieve mastery, however the hospital cannot provide it for all trainees in a planned, regular and uniform way; rather training is based on available opportunities. Under such circumstances a record of experience must be maintained in a systematic way, so that trainers and trainees can easily identify areas of weakness and remedial measures can be taken at the appropriate time. A logbook serves the important purpose of keeping track of training, but it does not record much about the quality of the performance, and does not provide an opportunity for writing down the reflections of the trainee on her/his performance. Keeping a logbook is not in itself an educational tool sufficient for learning and assessment.

Portfolios

The portfolio is not simply a list like the logbook, since it neither records all experiences, nor is it confined to recording of experiences. Portfolio preparation involves the collection of experiences, selection from this collection and reflection on the selected experiences. It is believed that these processes are thought provoking exercises, and hence a portfolio is seen as a tool for supporting development in technical, professional and personal competencies. Preparation of a portfolio provides an opportunity for self-assessment. This is valued educationally because it helps the trainees learn form their experiences. Reflection can be on anything from technical knowledge, management of resources and time to personal qualities.

What is in a portfolio?

A portfolio is a collection of materials documenting the educational experiences and training of a learner. It includes the trainees evaluation of the training and their reflections on their own performance.

The highly individual nature of a portfolio means that trainers should be free to decide its' structure. A properly prepared portfolio, that is discussed with peers and trainers, can identify the progress made by the trainee in areas such as planning, self-assessment and learning from experience. Although the experiences described in the portfolio are

The functions of a portfolio:

- A record of training experiences.
- An inventory of non-clinical experience.
- Raw material for making a C.V.
- A vehicle for reflection about clinical and non-clinical experience.
- A vehicle for reflection about progress.
- An assessment instrument for professional attitude and behaviour.

embedded in the context of training, reflection on those experiences can be used for monitoring personal development, rather than technical achievement. It might be argued that this approach would shift the focus from acquiring knowledge and skills to personal development, but the current belief is that personal and academic understanding are closely bound up with each other.[1]

Benefits from the process of preparing the portfolio

Educational theories supporting the use of a portfolio as a tool for learning suggest a number of benefits.

It promotes ipsative assessment

Portfolios are not useful for comparing the performance of trainees. They are best suited to comparing the performance of the trainee with the objectives of the curriculum i.e. portfolios support Criterion Referenced rather than Norm Referenced Assessment. The process of preparation of a portfolio requires the setting of standards of performance and then identifying the gaps between these and actual performance. Trainees can set targets, based on their existing level of performance, and then can strive to achieve this standard. In this way the process of maintaining the portfolio permits ipsative assessments. (An ipsative assessment compares a learner's current performance with his or her past performances)

It is qualitative in nature

Whether used for self-assessment or for formal assessment a portfolio reflects the quality of learning. Assessment with the help of portfolio uses estimates of overall performance and the outcome of the assessment is in the form of qualitative comments, useful for providing feedback for improvement.

Preparing a Portfolio helps develop:

- The ability to understand and take account of their own strengths and weaknesses when completing tasks or planning future learning.
- The ability to assess their own performance independently and in collaboration with others.
- The ability to learn from experience and apply that learning in new contexts.
- The ability to communicate effectively about their experience.

It assesses contextualised and situated knowledge

Since a portfolio is a reflection on a selected collection of experiences in real situations, it provides an opportunity to assess performance in real life settings. By being 'situated' or 'contextualised' in this way the portfolio provides the opportunity to assess other personal and professional qualities like leadership, communication and management skills that is not possible by traditional methods.

It provides 'Autonomy' and 'Motivation' for learning

The proper use of a portfolio may help in developing the trainees' independence as a learner and internal motivation. This step is crucial for life-long learning.

It is trainee centred, individualised and works as a 'mirror'

Since the design (what is to go in the portfolio), control (how much and when) and ownership (to be shared with whom) all rest in the hands of the trainee, it is a truly student centred educational tool that can cater for individual differences. The complete flexibility and freedom provided by a portfolio allows a range of learning styles to be used according to the preference of the learner. Most assessment methods are designed for assessment of trainees as a group, but a portfolio is inherently individualistic in nature.

The portfolio builds up an image of the trainee as a professional that reveals strengths and weakness. The trainee can see this image like looking at their reflection in a mirror. Importantly this image is the trainee's own creation. This is an important strength of the portfolio, since in other assessment systems, trainees often do not fully accept the trainers vision of them as true believing that it is coloured by prejudice and stems from faulty observations.

It works as a map of growth

The portfolio provides an opportunity for trainees to document their professional behaviour over time. Used in this way, the portfolio acquires importance as an instrument for recording development. This can be described as making a map of growth.

It helps in establishing a fruitful dialogue between trainees and trainers

If trainees (novices) share their reflection on their professional experiences with senior peers and trainers (experts), then a discussion can be generated to find out the solutions of the problems in handling complex tasks. These can centre upon the methods used by the 'experts' to handle complex tasks. Usually these processes are in the brains of the experts and remain hidden to novices.

It widens the curriculum

Setting the topics for assessment goes a long way towards determining the curriculum, because this is where learners will focus their attention. Traditional methods of assessment can assess only propositional knowledge (facts) and hence the focus is on learning in the cognitive domain. If the trainee knows that they will also be assessed on the breadth of material reported in their portfolio, and also on the nature of their reflective commentaries their learning efforts may be redirected towards these areas.

Some effects of the portfolio on the trainees' independence as learners:

- It makes trainees think about their own learning both at the technical level but also at the level of thinking about the way they learn.
- It may lead to critical autonomy by promoting reflection on the complexities of the profession and its effectiveness in serving the larger interests of society.
- It makes trainees compare their work with their peers and trainers.
- It may promote critical conversation between the trainees and trainers and amongst peers. This may lead to the trainees reviewing their own targets.
- It makes trainees talk and write about their own strengths and weaknesses, motivating them to improve.

The dilemma of use of portfolio for formal or summative assessment

The use of the portfolios for summative assessment (i.e. assessment for certification) is a contentious issue. Honest public reflection on performance requires that trainees do not feel threatened about any misuse of that reflection. It is natural for trainees to fear that they may reveal too much about their weaknesses and that this may affect their careers. If the portfolio is used for assessment by trainers it may undermine the true reflection on their performance by trainees. This concern may cause trainees to boast about their strengths rather than identifying their weaknesses. If some institutes or trainers want to use portfolios for formal assessment they must accept that the process of preparation may loose some of its educational value.

If the portfolio is used for assessment:

- The evidence presented must be valid; i.e. show what it is intended to show.
- The evidence must be sufficient; i.e. the quantity and variety of experiences and reflection on those experiences is detailed enough to ascertain that learning has taken place.

Some reasons why portfolios are used for assessment even though this reduces their value as a learning tool:

- "If we don't assess the portfolio most of the trainees are not going to fill one in." The spirit of portfolio is to promote autonomy in learning and how can be autonomy be promoted if the use of the portfolio itself is not autonomous!
- Some trainees may not be able to identify their own shortcomings unless their portfolio is reviewed with a trainer.
- The portfolio permits some assessment of non-clinical performance, behaviour and professionalism. These are so difficult to assess otherwise that it may be judged that the gain is worth the loss of portfolio development as a learning tool.

- The criteria for checking validity and sufficiency should be derived mainly from the actual context of the learning.

How to introduce and implement the portfolio supported learning and assessment

Some important considerations surrounding the introduction of a portfolio system:[2]

- *Portfolio based learning is not a soft option*: Most learners will find the process of developing a proper portfolio arduous. A great deal of thought and reflection goes into choosing and preparing the evidence of learning. This requires a consistency and candour that is unusual to find in other educational processes.
- *Portfolios are highly personal*: Individual learning preferences and pressures of work may dictate the amount of effort and the consistency with which the learner approaches the process. Trainers should recognise this and not impose a standardised format or dictate what will count as 'appropriate' evidence of learning.
- *Portfolios are about learning*: It should be clear to all that the portfolio is not a simple record of experiences in chronological order. The portfolio is useful only when there is reflection on experiences and subsequent modification of practice.
- *Quantity of evidence is no substitute for quality*: It should be very clear that a few examples of reflective learning in practice are more important than just a long list of the tasks performed.

The portfolio is not only a document, but is a complex and abstract concept. The process of development is much more important than the product. Both trainees and trainers need to understand this before designing and implementing any portfolio supported learning and assessment scheme. The following steps may be taken:

Orientation of trainees and trainers

Trainees will need to be taught something about portfolios and about their own learning processes. There will similarly need to be a full faculty development programme for the trainers.

Provide adequate support

In the beginning some trainees may find it difficult to engage in this process of preparation of the portfolio. Help must be given to these trainees if they want it.

Trainees may need help in two areas. Some trainees may find it difficult to reflect on their performance, and for such trainees trainers can provide probing questions based on the type of task on which they are reflecting. Some trainees may be able to reflect on their performance, but may find it difficult to express their views in writing. Such trainees may need some training in communication skills. However, if trainees are willing to discuss their performance with their peers and supervisors, such discussions may automatically generate a language that can be used to communicate their reflections. It may also prove helpful to supply some examples of properly and improperly prepared portfolios. If a lot of trainees are not able to start the process then it is advisable to conduct some workshops. In these workshops trainees may be helped to prepare portfolios for the tasks they do in short period (e.g. last week's experience). Such workshops may help trainees to appreciate the potential of the portfolio and will give them confidence that they can prepare it.

If the full potential of the portfolio is to be realised, trainees should be convinced that it is worth the time and effort they put in.

Provide time in the curriculum

No initiative can succeed (whatever are its benefits) if it is not provided with enough resources. This is true of portfolios. If there is no provision of time for preparation of the portfolio, then it will be seen as some extra imposition in an already busy schedule. It is advised that some time should be provided to the trainees exclusively for this purpose. This time can be as little as a half day every three months, but it will make a great difference. It is most important to provide time at the outset. After a while trainees will start feeling confident in this process and then they will be able to prepare the portfolio whenever and wherever time is available to them.

Because the portfolio can only be useful when it is prepared conscientiously and honestly trainers should ensure that trainees are aware of all of the possible benefits.

References

1. Gibbs G. *Creating a teaching profile*. 2nd ed. Bristol: Technical & Educational Services, 1992.
2. Challis M. AMEE medical Education Guide No 11 (revised): Portfolio-based learning and assessment in medical education. *Medical Education Medical Teacher*. 1999;21:370–385.

Section 2
Clinical teaching

10

Teaching anaesthesia in the operating theatre

David Greaves

Operating theatres are good places to teach anaesthesia because the learning is:

- Active
- Experiential
- In a real context
- Repetitive
- Problem orientated
- Encourages the adoption of role models
- Makes economic use of resources.

Anaesthesia is learned in the course of real clinical work, and though clinical teaching is an essential element of medical education relatively little effort has been put into understanding it. Anaesthetists use the operating theatre to learn in three ways:

- It is the place where they learn the practice of anaesthesia. This includes practical skills and procedures, how to look after and monitor patients and how to make clinical decisions.
- It is a place where the theory of clinical science can be connected to real events.
- It is the place where they learn how to behave as professionals.

In-theatre teaching should address all these aspects of learning.

Supervised learning in the operating theatre is active, experiential, problem orientated, task based and repetitive. In experiential learning the trainee takes part in the process about which he is learning, and this practical context is believed to help both

The activities of anaesthetic teachers:

Supervising

- Keeping the patients safe.
- Preventing the trainee from taking potentially dangerous decisions through ignorance.
- Using expert skills to rescue situations that are beyond the capabilities of the trainee.
- Managing the work environment.

Teaching

- Showing the trainee what to do.
- Guiding the trainee's clinical decision making process.
- Directing the trainee's attention.
- Acting as a role model.
- Determining learning objectives.
- Developing problem solving skills.
- Discussing the principles on which the science of anaesthesiology depends.
- Developing learning situations.
- Monitoring the trainee's progress in order to plan their further experience.
- Assessing competence.
- Checking that learning has occurred.
- Encouraging reflection.
- Encouraging self-assessment.
- Giving feedback.

understanding and recall. Working in the operating theatre provides a good opportunity for problem-orientated learning. The trainee uses his clinical reasoning skills and the consultant acts as moderator, helping with decisions whilst keeping the patient safe. Problem orientated learning must be distinguished from Problem Based Learning. (See Chapter 7.)

The learning is taking place in a vivid context, is active and repetitive. Repetition is required both to reinforce the learning of motor skills and in helping to remember what to do.

The theatre as a classroom

The operating theatre is organised for its' primary purpose of performing surgical operations and this interferes with teaching. Theatre work, and the encounter between trainee and consultant that it provides, should be seen as the main focus for anaesthetic education. Consultants who take the trouble to teach their trainees how to behave in theatre and set them a good example, are providing a lesson that cannot be learned anywhere else. In other respects, however, the theatre is not a good classroom. Talking to the trainee in theatre may distract and annoy the surgeon, particularly as operating theatres are already noisy from air conditioning, suckers, diathermy, monitors and background music. Conversely, these noises are a distraction from teaching. Face-masks hide expression, and the teacher and student are often prevented from facing one another by the placement of patient and equipment.[1] All these factors hinder communication.

Trainees and consultants will share the practical tasks on an operating list providing repeated cycles of demonstration and performance.

A trainee anaesthetist is continuously challenged by the demands of providing anaesthesia. Learning may be the last thing on their mind as they focus on the issues of professional survival.[2,3] Byrne has shown, in a study using the Access simulator, that as tasks become more complicated errors in record keeping occur, demonstrating that the routine tasks of anaesthesia leave the trainee with little spare capacity to attend to teaching.[4] One of the skills of teaching in theatre is providing the trainee with the right load for learning, and recognising when he or she is too busy to pay attention.

An operating theatre is a stressful workplace.[5] Mistakes, even small ones, can endanger the patient's life, so surgeons and anaesthetists are often very tense. The work is emotionally charged even for those used to the environment. For those new to operating theatres the personal stresses can be very high. Operating theatres are expensive to run and everyone who works there must co-operate to get the days work done. The work scheduled frequently exceeds the time available and emergencies commonly interfere with routine.

As the consultant works and teaches, the trainee is learning as much from how they act as from what they say. Trainees can learn through experience, by instruction and by following the consultant's example. Learning by example, providing role models and shaping behaviour is crucial to the professional development of doctors and can only take place in real clinical situations.

The learner must feel safe

The learner must feel safe:

- The learner must understand what is expected of him.
- The learner must be capable of the work demanded of him.
- The consultant must protect the learner from harassment by other members of the theatre team.
- The trainee must not be afraid of their supervising consultant.
- The learner must have sufficient time to do things.
- The learner must know that competent help is always at hand.
- The learner must not feel that they are letting the team down.

It is the responsibility of a consultant who teaches a learner in the operating theatre to provide them with a safe and secure environment in which to develop their skills.

Deciding who is in charge

If you are going to teach your trainee on a topic by discussion in theatre:
- Take over the primary responsibility for the patient.
- Take over responsibility for the anaesthetic record.
- Position yourself so that you have a good view of the patient and monitors.
- Pay careful attention to the patient as you teach.

When two anaesthetists work together in the operating theatre they must agree the division of responsibilities. It is important that one anaesthetist has clear responsibility for the patient, and that this responsibility is passed from trainee to consultant, and vice versa, formally, so that there is no doubt in anyone's mind. This is sound clinical sense, as there are many instances of patients coming to harm because each anaesthetist thought the other was caring for the patient. It also makes good sense for education. Trainees feel responsible for the patient, even when they are working with consultants. They are usually very vigilant for the patient's safety, and if they feel that the consultant is neglecting the patient to teach they will become agitated. They will then not concentrate on what is being said. When the consultant is going to teach he must always take over control of the patient. She should say something such as:

Who is in charge?
Only one anaesthetist can take the lead in giving an anaesthetic and both the consultant and trainee must know who is running things.

"I am going to talk to you about ... and I will monitor the patient so you can listen."

She should then position herself so she can see the monitors clearly and take over the recording of the anaesthetic. The consultant should continue to attend to the patient and make clear responses to events. She should continue to keep a good anaesthetic record. The learner will usually be reassured and will attend to what the consultant has to say. Some learners are unable to attend to teaching during anaesthesia. The consultant must be on the lookout for this trait and must respect it. Such doctors will probably also be unable in turn to teach in theatre and this must be taken into account as their career develops.

What should the trainee know about learning in the workplace?

For trainees to get maximum benefit from theatre teaching they must understand the full range of what they can learn, including matters such as personal conduct and professional behaviour. They should also understand the cycles of experience and explication, and the value of reflection.

Medical students on ward rounds more often report that they have been taught if the teacher prefixes each episode of teaching with a message that what is to follow is teaching. In theatre the trainee often does not have a clear idea of their role, either as a worker or as a learner. A little explanation can help them. In a hospital with a particular method and style of teaching it is useful for trainees to understand how the teaching is supposed to work, and how they are intended to learn from it.

The consultant must remember that the trainee is learning how to behave by watching how *they* behave and by copying what they do. The most potent lesson in the course of real work is the behaviour of the consultant role model — beware!

Trainees will behave in theatre in the same way as their teachers.

Giving structure to operating theatre teaching

A learning objective is a defined learning task, agreed upon by teacher and learner and providing a purpose for an episode of teaching. It may be assessed by observing its outcome.

Teaching in the operating theatre is a formal educational encounter and should have clear educational objectives. That is, teaching in the operating theatre should be planned. The theme should be followed during the course of the practical work and at the end of the session there should be a review of the topic. The key to successful teaching is the ability to lend structure to the day's experiences, whilst remaining flexible enough to capitalise on unexpected opportunities. Little research has been done into clinical teaching techniques and how best to plan bedside and workplace education. It is often thought that the unpredictable nature of work makes teaching in clinical settings equally unstructured. Clinical teaching is the best way to teach some parts of the curriculum, and consultants need to know what they can most profitably teach in theatre. Working in the operating theatre is not only an opportunity for acquiring knowledge, gaining understanding and learning skills, but also an opportunity to learn about attitudes, communication and other affective aspects of practice.

Getting ready to teach

There are three phases in a theatre teaching session:

- Getting ready to teach
- Doing the work
- Reviewing the session

Traditionally anaesthetists visit their patients before surgery, to make an assessment and discuss the forthcoming procedure. This practice has been affected by increasing numbers of day stay patients, the admission of patients on the morning of their

Before the list:

- Present a summary of each patient's relevant history.
- List the particular problems which each patient presents.
- List any likely complications.
- Describe an anaesthetic plan for each patient. Where necessary the trainee should be asked to give detailed explanation of their planning.
- Identify any other problems and difficulties relating to running the list.
- Take particular note of the trainee's plans for each case in order to observe and ask about any changes to these plans.

During the list:

- Observe the trainee at work. Try not to interfere or distract too much.
- Let the trainee plan and conduct anaesthesia.
- Ask the trainee to explain his or her thinking as they go along.
- Periodically ask the trainee to evaluate what is happening.
- Let the trainee make decisions about list organisation etc. Do not respond if other staff involve you in the anaesthesia.
- Watch the conduct of anaesthesia carefully at key times. Times such as first incision, cross clamping, delivery of the baby etc.

Note examples of particularly good or bad decision making, skill or behaviour. The trainee should be asked to jot down things that they are pleased with because they went well and things that they regret because they went badly.

After the list:

- Take five minutes to discuss the days work and round off.
- Ask the trainee how they feel the day has gone.
- Talk about their list of the things they think went well and the things they think went badly. Try to let the trainee do most of the talking.
- Give praise where it is due.
- Never offer a criticism without suggesting how matters might have been better managed.
- If there are instances where the trainee departed from his original plan ask what caused the change of mind.
- Discuss your list of things that went well and things that went badly.
- Achieve a consensus on how the day has gone.
- Suggest any suitable activities to follow up or read. Always try and suggest something that the trainee should reflect on.

operation and the use of pre-admission clinics. The work arrangements of trainee anaesthetists may also preclude them visiting the patients on the day before surgery. In some anaesthetic services, formal systems for peri-operative management have superseded pre-operative visiting. It is still reasonable, however, to expect that a trainee will familiarise themselves with the problems of the patients they will anaesthetise on teaching lists! Novice trainees must be taught how to conduct pre-operative assessments by attending them on the ward or in the clinic, along with the consultant. Trainees must become familiar with the protocols being used to plan pre-operative investigations, and the way the anaesthetist gets to know about the patient's general health.

Some factors influencing a teaching plan:

- The trainee's personal learning needs.
- The trainees degree of experience.
- The type of work to be undertaken.
- Any pre-existing problems with the patients.
- Any specific requirements of the surgery.
- Any special experience appropriate for the trainee.

No matter how pre-operative assessment is done, at some time before the start of work the consultant and trainee should spend a few minutes discussing the days work and setting learning targets for the list. This constitutes the briefing, and is directed towards setting the educational framework for the clinical teaching that will follow. Briefing is the means by which the learning that is to come can be linked to what has been learned before. The consultant will want to know the extent of the trainee's experience both in general, and in relation to the particular work being planned. Consultants should do their best to find ways to develop worthwhile lessons from the day's work. A good approach can make quite mundane work interesting.

A contentious issue for consultants is how to deal with the trainee who arrives in theatre without having familiarised themself with the patients. Some consultants will not work with a trainee who comes unprepared. How should this problem be approached? In the first place the consultant must decide whether the trainee has a good reason for their omission, or whether it shows a habitually slipshod approach. In this latter case, they should remind the trainee that, without adequate knowledge of the patient, the condition and the surgery, they are not equipped to make decisions. They should tell the trainee that their attitude suggests that they might behave similarly when they are solely responsible for patients, and this is unacceptable. They should tell them that, by their lack of background knowledge of the patient, they have relegated themselves to the role of assistant and spectator. If the consultant does not need them as an assistant they should be sent to see the other patients. It is this authors personal opinion that the consultant should never become angry and 'tell off' the trainee. Such a response makes further learning unlikely.

Trainees have their own learning needs

Trainees often come to a list with a strong need to learn something specific, and it is better to work with this, rather than to try to impose new targets. The main

"The anaesthetist called in sick. What's your poison?"

preoccupations of trainees are examinations and coping with independent practice. The consultant should find out whether the trainee is close to taking an exam and take note of this in the emphasis placed on the day's work. The need to cope in independent practice is very important and provides a strong motive to learn. The early learning of an anaesthetic trainee is directed towards establishing a safe sequence of techniques and developing a means for deciding between them. As they learn, they relate new techniques to this framework, and then incorporate them. The stimulus of new challenges ensures that the learner is always seeking to enlarge their repertoire. When planning a theatre teaching session the consultant should be at pains to explain to the trainee where he can use what he is learning, and why it will be of use to him. If it can be shown to resolve a difficulty that he is having, then so much the better. Shysh showed that in his hospital neither trainees nor consultants thought they gave enough consideration to the trainees' specific learning needs for theatre teaching.[2]

Doing the work – watching and helping

During the list the consultant will watch and help the trainee. The consultant must concentrate upon making his educational plan come to life. This will require that the trainee has a suitable degree of autonomy, but that the consultant is sufficiently involved with events to be able to direct and control the trainee's performance.

Allow the trainee to take responsibility

It is not uncommon for trainees to be denied all experience of being in charge by the presence of a consultant. In some cases the consultant does everything that is of critical importance, and the trainee becomes an observer, robbed of all chance of making clinical decisions. Often the trainee is more subtly relegated to a passive role, being nominally in charge, whilst at all critical times the body language and attitude of the consultant make clear that *he* is the real leader. Consultants should let their trainees direct the theatre team, organise patient positioning, monitoring and give the surgeon permission to start. They should deflect questions put directly to them by surgeons and theatre staff. This delegation of authority should be as passive as possible. Passing a request on to 'the lad' with a flourish is a potent expression of power and status.

The consultant and trainee must agree between themselves what level of responsibility the trainee is accepting, and performing to that level must be one of the lessons for the day.

Hints for trainees:

- Trainees can help the consultants to organise their teaching.
- If the consultant doesn't have a plan ask him to make one.
- If you have a plan suggest it to the consultant.
- If you haven't got a plan ask the consultant what he or she thinks are the best things for you to do.
- Make a note of the things you think went particularly well or particularly badly and ask the consultant for comments.
- Ask the consultant to define your responsibilities for each case.
- Be clear about whether you have the primary responsibility for the case.

Hints for trainees:

- Negotiate your level of supervision.
- If you want to do a particular case then suggest this to the consultant and negotiate an appropriate level of supervision.
- If you have doubts about some aspect of your performance ask the consultant to watch you and provide appropriate advice.
- Always pay attention to the needs of the rest of the team.
- Ask how things went at the end of the list.
- Ask the consultant how he thinks you are doing and about your readiness for independent practice in this area.

Let your teaching respond flexibly to events

Whilst advising consultants of the importance of making their theatre teaching planned and structured, the point must be made that the teaching for a theatre list can be planned in too much detail. Surgery and anaesthesia are unpredictable, and the teaching must be flexible enough to take advantage of unexpected opportunities. If the full advantage of situated learning is to be seen, then use must be made of memorable contexts as they arise.

References

1. Paget NS, Lambert TF. Tutor–student interaction in the operating theatre. *Anaesthesia and Intensive Care*. 1976;4:301–330.
2. Shysh AJ. *The nature of anaesthesia residency education in the operating room at the University of Calgary* (Master of Science dissertation). Alberta: University of Calgary, 1997.
3. Gaba DM, Lee T. Measuring the work load of the anaesthesiologist. *Anesthesia and Analgesia*. 1990;71:354–361.
4. Byrne AJ, Jones JG. Errors on anaesthetic record charts as a measure of anaesthetic performance during simulated critical incidents. *British Journal of Anaesthesia*. 1997;80:58–62.
5. Gaba DM, Howard SK, Jump B. Production pressure in the work environment. *Anesthesiology*. 1994;81:488–500.

11

Helping trainees to develop decision-making skills

David Greaves and Michael Olympio

> *Clinical teachers should help learners to develop the habits of mind that will help them solve problems.*

Decision-making is one of the main non-technical skills of anaesthetists. It is surprising that clinical teaching does not usually include formal instruction about how to make decisions and solve problems. Consultants demonstrate practice, guide trainees through procedures and give mini-tutorials, but they do not often specifically instruct them on how to approach a clinical puzzle.

At the start of an anaesthetic training the novice anaesthetist is confronted by a bewildering number of problems. Many trainees try to learn a repertoire of responses to each situation. Inevitably, as the complexity of their work increases, they construct more rules than they can keep track of. They must be helped to learn to take a more flexible approach to decision making, and to search for solutions to each problem individually.

Unfortunately there is no clear definition of clinical-decision-making skills, and there is little understanding of how we learn and use them. Clinical decision-making

The key non-technical skills:

- Situation awareness
- Decision making
- Task management
- Team working

has been examined in the context of the diagnostic consultation, but not in clinical anaesthesia. Studies in internal medicine have shown that experts use problem solving approaches that are very individualistic, and that they home in on the correct decisions more efficiently than novices. There is controversy between those experts who think that doctors reason by logic (either from cause to effect or by an algorithmic series of small decisions), and those who believe they have an intuitive approach to problems. Michael Polanyi made the observation that although experts use a repertoire of knowledge and skills, they frequently cannot explain how they make their decisions.[1] He postulated that this expert process came from the use of hidden or 'tacit' knowledge. Tacit knowledge can only be gained from experience. Expert decisions are not solely, or even mostly, based on tacit knowledge. Experts also have a fund of explicit skills and knowledge that they have worked hard to master. It is important to realise that, though it may be difficult to explain how experts have arrived at their best intuitive decisions, inexplicability is not a necessary feature of expert decisions. Some clinicians make the mistake of believing that an irrational leap to a decision must be right, because it is inexplicable.

It is difficult to understand how experts arrive at their decisions.

Elstein and Schwarz have described two competing groups of theories about how doctors arrive at decisions.[2] Problem solvers believe that hypothesis formation is at the root of the process. The available evidence is used to form a hypothesis, that is then the basis for the search for additional evidence in a hypothetico-deductive process. Expert problem solvers are found to have great experience in their field and appear to work by comparing clinical pictures with prior instances of similar cases. They do not believe that the cases are the same, but seem to use their nearest match for a process of hypothesis refinement.

This introduces the second group of theories; those rooted in a theory of decision making. Adherents of this view of clinical practice believe that doctors use evidence to weigh the choices between competing possibilities. This approach requires knowledge of what the chances are of one hypothesis being more likely than another. Bayes's theorem can be used to combine the mathematical probabilities and calculate the most likely answer. The 'decision-making' route to making a choice is attractive to many doctors, as it provides a link between evidence based medicine and clinical reasoning, because the application of Bayes's theorem requires accurate knowledge of probabilities. Proponents of the more cognitive based 'problem solving' approach to practice believe that the available evidence should be used largely to confirm the likely efficacy of a course of action, rather than to mandate the reasoning process towards it. There is not much published evidence about the decisions of anaesthetists in the fast moving world of the operating theatre. There is no tradition of recording and publishing reports of how the monitored observations change, and what presentations are most frequent in terms, for instance, of a condition such as sudden, unexplained hypoxia. How does the precise clinical picture of pneumothorax differ from that of severe bronchospasm? How soon is the trachea found to be deviated? Does hypotension, usually precede hypoxia, or vice versa. At present we have no statistics to apply to Bayesian logic.

These uncertainties open the way for the enthusiasts for routines and protocols. Here the deductive process relative to the pathology is subordinated to a fixed schedule of actions, that are believed to have the best outcome for a mixed population of patients. Such protocols usually incorporate a number of if/else choices, and are constructed as algorithms. In some conditions such approaches are very useful, but in others the choice between limbs of the algorithm amounts to a difficult clinical choice, and we are back where we started.

What are we to conclude? Clinicians must reason carefully, make deductions from the evidence, using what they know, know the probabilities that have been determined from evidence based medicine and make judicious use of conventional algorithms, where such exist!

It is probable that there are some behaviours that are necessary for clinical reasoning. Educational psychologists have postulated a 'tool-kit' needed for rational decision-making, and it is possible to prompt trainees to use these apparently generic skills. A teacher can observe and ask questions that direct the trainee towards them. The aim is to cultivate habits of mind, so that they will think in similar ways when they work alone. Learning is best accomplished when learners understand what is expected of them, and they are often too involved in the events to be able to analyse their own decision making without prompting. The consultant needs to help the learner critically reflect on the way they are making decisions.

Setting the scene for teaching decision making

Trainees always learn best in a supportive environment, but this is particularly important when the consultant is actively teaching the techniques of decision-making. Many of the questions which will arise (Why did you do that? What else could you have done?) sound like criticism, and carry an element of threat. Working with a consultant who constantly questions the trainee's decision can raise a high level of anxiety. One way to reduce the level of threat is for the consultant to explain what he or she is doing, and describe to the trainee how the skills of clinical decision-making can be learned.

Is the trainee ready to learn about decision making?

The trainee will be ideally prepared to learn if they come to the operating theatre knowing how the consultant will be guiding them to learn decision-making skills. He or she must understand that learning is a questing, active process. As Willenkin has put it:

"In order for effective learning to take place, learners must accept their need to not just take in information and regurgitate it on request; but to assimilate it, process it, reorganise it, generate new ideas from it, connect new information to previously

learned material, understand the meaning and value of new information, think about how and when to use this data, and determine how new data answers old questions and raises new ones."[3]

It is this willingness to learn and explore that makes the difference between teaching a passive unresponsive trainee and teaching one who is lively and challenging.

The learner must also come to learn with a sense of personal responsibility. Learning is always ultimately the individual's own responsibility, and only a trainee who has accepted this will be able to learn effectively. In addition they must be committed to caring for the patient as they are learning. This involves being prepared to balance their needs against the rights and requirements of the patient.

Consultants as teachers and role-models

The consultant role-model should provide a secure environment:

- Project confidence *and* humility.
- Be respectful and receptive to others.
- Allow acceptable alternatives.
- Be professionally assertive; *earn* your respect.
- Use eye contact.
- Defer (appropriate) answers to the trainee.
- Provide graded responsibilities.
- Ensure the safety of the patient.
- Be constructive: include the surgeon in your decision process.

Trainees must respect their consultants as teachers and as doctors in order to learn decision-making from them. If a trainee thinks that one of his teachers is a fool, with nothing useful to say, they are not likely to learn much. Consultants must understand that trainees will only respect them as teachers if they earn that respect.

Sometimes trainees are frightened of their teachers and will not learn at all in this situation. Some otherwise sensitive consultants do not comprehend that learning about decision-making is a personal threat to the trainees self-esteem and confidence.

Consultants who hope to teach a rational approach to decision-making should strive for excellence themselves. Epstein observed that exemplary physicians have the capacity for internal reflection and self-awareness of their skills, limitations and value systems.[4] They constantly compare their performance with their internalised standards, and strive to achieve their own goals. This self-reflective criticism pervades every aspect of their professional life, including their interactions with patients, colleagues, mentors and trainees. Mindful practitioners seem to have well-developed value systems and strong affective skills. Exemplary physicians describe their use of peripheral vision and heightened awareness to discover obscure variations, and a natural curiosity to ask questions about those findings. They consistently bring 'tacit' (experiential) knowledge to the forefront, continuing to add structure upon a solid foundation. They lay aside prejudices and see things as they are, and not as they would like. In solving problems they exhibit humility to question their own judgement, to ask for help, and to self correct their technical skills. These are the qualities of practice that the consultant hopes to demonstrate to the trainee.

Good trainees have the same qualities and these consistently help them to struggle through arduous training programs with commitment and responsibility. Having this mindset enables the trainee to look for abnormal findings, and compels them to solve problems. They aspire to the same qualities of practice that are shown by the best amongst their trainers.

Give the trainee the responsibility for decision-making

You can't learn how to make decisions without making decisions! A prerequisite of teaching decision-making in the operating theatre is that the trainee is given the responsibility to make decisions, and to act on them. Consultants should try to maintain a consistent and appropriate level of involvement in the anaesthesia. It is very hard to stand back and watch, when the trainee is working at the limit of their ability, but it is in this situation that maximum learning is taking place. Trainees usually defer to their supervisors, and if the consultant begins to interfere the trainee will stop deciding what to do and wait to be told. Consultants can inhibit their trainees from making decisions by appearing anxious. They may show this by their body language without saying a word. The teacher who wants the learner to make decisions will give them control of the anaesthesia, and then be as unobtrusive as possible. Consultants are often very anxious about the actions of the trainees they supervise, and must learn to understand their own fears and not step in unnecessarily.

Clinical decision-making can be a life and death matter to the patient and a heavy responsibility for the doctor. Watching and offering a commentary is far less involving, and it is probable that such thinking does not emulate real clinical problem solving. Only the burden of real responsibility will complete the mental apparatus required for clinical judgements. The consultant has the difficult task of supporting the trainees practice to keep the patient safe, whilst allowing them enough room to feel responsibility for decisions.

Doctors approach clinical reasoning tasks in their own way, and supervisors must allow trainees some latitude in their problem solving. A consultant who insists upon rigid observance of his own methods is not allowing the trainee room to manoeuvre and learn. It is, for instance, likely that a trainee will be slower to observe and respond than an experienced consultant. They will also have a tendency to act in response to stimuli that the experienced anaesthetist would ignore. The learning process involves experiment, within the limits of acceptability.

What is the taxonomy of learning objectives and does it matter?

A taxonomy is a classification. Learning objectives have been classified by Bloom into those relating to knowledge (cognitive), skill (psycho-motor) and behaviour (affective).[5] Each of these domains is subdivided into a cascade of functions in order of their complexity.[6] The taxonomy that has yielded the best practical insight in organising teaching is that relating to the cognitive domain.

Taxonomic categories of educational objectives in the cognitive domain.[5]

	Definition	Example of objective
Knowledge	The remembering of previously learned material. All that is required is the bringing to mind of the appropriate material. This is the lowest level of the cognitive domain.	"Learn a list of the causes of sudden hypotension during surgery."
Comprehension	Recalling and grasping the meaning of the material learned. This is exemplified by the capacity to summarise the material.	"What do you mean by hypotension during surgery."
Application	The ability to use the learned material and to know what the outcome will be. This is a higher cognitive skill than comprehension	"Record the blood pressure and tell me when the patient is hypotensive."
Analysis	Analysis is the breaking down of learned material so that its organisational structure can be understood. This involves the identification of parts, the relationship between parts and the way the parts are linked together. This is a higher cognitive level because it requires knowledge of the internal structure of learned material.	"Classify the causes of hypotension during surgery."
Synthesis	Synthesis is the ability to put the component parts of knowledge together in new ways to make a new whole. Learning outcomes dependent on synthesis skills are creative.	"Learn about how we can treat different types of hypotension occurring during surgery."
Evaluation	Evaluation involves judging the worth of learned information. It includes the ability to compare, discriminate an allocate to hierarchies.	"What is the reliability of your observations of blood pressure."

Decision-making is largely, though not exclusively, the exercise of abilities in the cognitive domain. Within this field of knowing and understanding learned abilities are ranked from the simplest to the most complex.

Keep your teaching at one taxonomic level

It is important to be aware of these levels of learning functions when teaching in the operating theatre, because trainees find it distracting to be asked to answer questions that require them to move about between the levels of understanding. Questions that require the trainee to list knowledge (a low-level function), whilst simultaneously evaluating and judging that information, cause confusion. The answers given will frequently be entirely knowledge-based, or entirely evaluative in nature and misunderstandings between teacher and learner can degenerate into the game of 'guess what I'm thinking?' Good teaching and good testing will keep questions requiring different levels of understanding separated.

If the trainee is engaged in administering anaesthesia in the operating room, their mind is, one hopes, working at the levels of analysis, synthesis and evaluation. Questions that focus their thinking at this level will be dealt with readily. A trainee will willingly respond to invitations to explain the theoretical basis of what he is doing. He will stumble and become irritated if asked to list knowledge, or if he is asked to evaluate and understand some matter that is unrelated to the matter in hand. The teacher should not distract the trainee by asking them to consider issues that require attention at a different taxonomic level to the one they are using. The objective of the consultant must be to prompt the learner to start thinking in the terms that are useful in solving problems.

Teaching decision making skills

In order to understand the trainee's decisions the consultant needs to know:

- What the trainee knows.
- What the trainee is observing.

Decision-making is one of the four, key, non-technical skills of anaesthetists. It relies on another of these skills: situation awareness and operates through the other two: task management and team working. In order for the supervising consultant to understand the trainee's management of the patient they must have an appreciation of the frame in which the trainee is working. As the consultant and trainee work together, their clinical reasoning and decision-making skills run in parallel. If the trainee is not responding to events, or is making decisions that the consultant cannot understand, the consultant will begin to feel uneasy. The supervisor does not consciously experience this dissonance. Rather it rises from within, resulting in irritation and disquiet. This is then attributed to the trainee's incompetence! The consultant must analyse the trainee's performance and decide what it is that is causing this uneasiness.

Hints for trainees:

- Learn the routines for problem solving and practice them yourself.
- Try always to anticipate the outcomes of your decisions and be on the look out for consequences of your past decisions.
- If you are working with a consultant who doesn't understand these routines ask him or her about relevant decisions. (Why? When? What do you think will happen? Are any of these issues connected?)

As the work proceeds the trainee should be posing questions to themselves and looking for answers. The activity of the consultant is in two parts. They should watch the trainee at work and, at appropriate times, ask relevant questions that address the various stages of decision making behaviour. Consultants are apt to wait until clear decisions are presenting themselves and then discuss the decision. Rather than doing this, they should from time to time ask appropriate questions to try to understand how the trainee is approaching the problems.

How does the trainee work?

A number of attempts have been made to list and prioritise the skills that anaesthetists use during clinical work. There is agreement that a preliminary process of orientation, called here situation awareness, is followed by the decision making process.

Situation awareness

Collection of information

Situation awareness has been discussed at more length in chapter six. There are three principle components. The first of these is 'collection of information'. This takes in a number of qualities of practice, such as perception and vigilance. Is the trainee seeing the important events that are occurring? It is important to know what they have seen, what they ignore and which observations are influencing their actions. Perhaps the trainee has noticed something and is not responding to it because he or she knows that it is not important. Perhaps they haven't noticed it at all.

Lack of vigilance differs from failure to perceive in that the trainee is not even looking! Inattention is a serious problem in an anaesthetist, and it must be dealt with. The consultant can get some idea of what the trainee is taking note of by:

- Identifying a problem for consideration.
- Directing the trainee's attention towards the problem.
- Getting him or her to describe the problem.

The three main components of situation awareness:

- Collection of information.
- Assimilation of information.
- Projection of what will happen.

When you have identified a problem, get the trainee to tell you when it started, whether it has been constant since they first noticed it and whether there are any other observations associated with it. The trainee's actions should be based on observation. If the trainee starts doing something, you can ask them what they have seen that has provoked them into action. If you have seen something that you feel they should have seen, and should be acting on, try and draw their attention to it indirectly. If you see the penny drop, and they leap into action, you may conclude that they had not noticed what was going on. Ask them if anything they have seen suggested that they might take a course of action, and why they decided against it.

A trainee's observations can sometimes be made clear by asking them to make a commentary as they go along. If the trainee seems to be ignoring important information, then they should be asked what they are observing. This will reveal to the teacher where the trainee is concentrating their attention. They should then be asked to explain what their observations mean. This will enable the consultant to understand whether the trainee understands that what she is seeing is a problem. The trainee may see and understand the problem and have decided not to take action.

Assimilation of information

This is the second step in situation awareness. What does the trainee understand by the information they have gathered? You need to know what significance the trainee is attaching to the events they see. Questions should be directed towards the trainees understanding of:

- The nature of the issues raised by an event.
- The urgency of the problem.
- The degree of threat.
- The ease of resolution.
- The need for action.
- Possible actions.
- Possible outcomes.

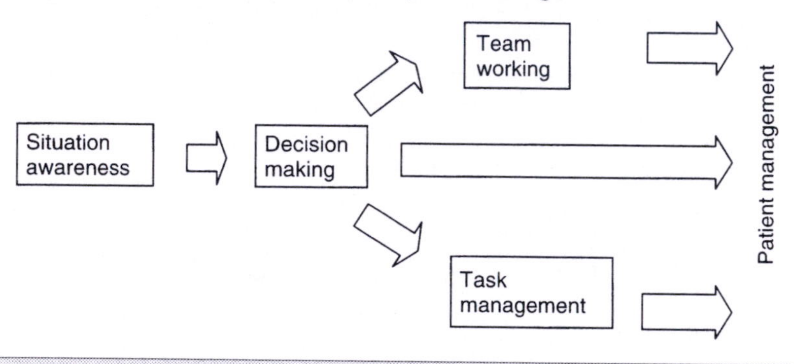

A problem solving sequence to use when teaching:

- Identify a problem for consideration.
- Direct the trainee's attention towards the problem.
- Get him or her to describe the problem.
- Analyse his or her understanding with regard to:
  - Comprehension of the issue.
  - Urgency of the problem.
  - Degree of threat.
  - Ease of resolution.
  - Need for action.
  - Possible actions.
  - Possible outcomes.
- Direct the trainee's attention to questions where his analysis or behaviour seems questionable:
  - Deficient knowledge.
  - Failure to perceive.
  - Failure to understand what is seen.
  - Failure to evaluate what is seen.
  - Failure to formulate a satisfactory plan.
- Review your assessment of the episode of problem solving.
- Use appropriate prompts to direct the trainee's behaviour.
- Give the trainee feedback.

There is evidence to support the view that learners may hold 'incorrect' opinions about phenomena, processes and their relationships, and that they may continue to sustain these opinions in the face of evidence of a gap between their model and their observations. In these circumstances the trainee may either ignore the correct explanation or, against all logic, hold contradictory explanations simultaneously. In this case, anaesthetic trainees may not learn very much in theatre, but simply generate rote-learned solutions to clinical problems. They keep their incorrect model and their learned solutions separate, and theatre experience does not improve their understanding.

Willenkin has suggested a variety of questions that the consultant can put in order to reveal the trainees thinking.[3] These can be used as 'teaching scripts' for the dialogue. He suggests that the approach should be essentially Socratic. The consultant may outline a principal, point out a fact or make a demonstration as a starting point. Thereafter the Teacher should use appropriate questions to draw out the trainee's insights and weaknesses. The consultant demonstrates and models each element of the decision-making process, giving the trainee repeated opportunities to demonstrate a similar process in return. The teacher will provide constructive criticism reinforcing proper behaviours and discouraging improper ones.

Projection of what will happen

The final component of situation awareness is to be able to project future events. How are things going to turn out? The good practitioner will constantly be anticipating problems before they develop. They will understand the consequence of actions and events and will be able to relate current events to previous occurrences. If the trainee does not appear to be looking ahead, then they should be asked questions requiring prediction and anticipation. If they appear to have a rigid approach, questions can be asked about alternatives. This process of directed development is much more effective than the critical approach where the consultant makes negative judgements:

"You are not responding to important data", "You don't anticipate" and *"You are too rigid".*

The consultant will also need to ask questions to see if the trainee understands the origins of a current problem. In particular they should be asked about alternative actions they could have taken, and how these would have impacted on the present situation.

Aspects of competent practice:

- Knowledge
- Skill
- Perceptiveness
- Confidence
- Prudence
- Vigilance
- Fluency
- Decisiveness
- Anticipation
- Organisation
- Flexibility
- Responsiveness
- Manner
- Assertiveness
- Management
- Communication

Task management

Good approaches to problem solving require that the trainee has organised their thinking. Do they have an overall plan? Are they able to integrate their current problem with the overall conduct of the anaesthetic?

Cleave-Hogg and Benedict's scheme for examining how a trainee is solving problems:

- Identify the trainee's basic level of knowledge.
- Help the trainee develop a theatre action/organisation plan.
- Begin with the current medical case and review the resident's organisational plan.
- Challenge the resident to be prepared for the unexpected.
- Direct the resident to reflect on organisation and implementation of the plan.
- Provide immediate and honest feedback in a constructive manner.

Trainees who are doing well with all aspects of their decision-making should be asked from time to time to justify their decisions, compare their actions with alternatives and evaluate the status of the anaesthetic.

When the teacher does a case with the trainee watching he should make his thinking as explicit as possible, and ask questions about how he should be behaving: "What would you do next, if you were me?"

The trainee should also be able to organise the steps of managing the investigation and treatment of a problem. This will include organising staff and equipment and co-ordinating anaesthesia with the activities of the surgeon. Consultants should watch how well their trainees undertake these elements of decision-making and give prompt feedback. They should also make sure that they model these skills when they are taking the lead.

Teachers should build for themselves a range of 'teaching scripts' that are ways of bringing the trainees attention to the particular aspects of decision making that they are finding difficult.

Cleave-Hogg and Benedict in their ethnographic analysis of the beliefs of good clinical teachers discovered that they used both Socratic enquiry and apprenticeship models of teaching.[7] In enquiry mode they probed the beliefs and understandings of trainees with probing questions. Apprenticeship encouraged learning through observation, experience and copying the actions of the consultant.

Some ways of focussing on decision making

Going with the flow

Make use of the random flow of problems that the events present. The consultant can direct the trainee's attention towards the aspects of problem solving that he feels are

Some techniques for exploring problem solving: Examine alternatives.

- Pick on a decision the trainee has made.
- Call the trainee's attention to the decision they have made. Is there general agreement about airway management for these cases? What sort of cardiovascular problems tend to arise during squint surgery?
- Ask the trainee what has influenced their decision and make them address a variety of factors. (The knowledge they have, the circumstances of the patient, the nature of the surgery.)
- Ask whether anything they have observed or any previous decision they took has influenced their current judgement.
- Ask them what problems may arise as a consequence of the choice they have made and how and when we will see evidence of these problems.
- Ask what different problems would have arisen if we had adopted a different plan.

Element of judgement	Appropriate question
Analyses and organises data. Knows what is currently relevant	What data are you taking note of at present? What are you thinking about? What are the current problems that you can see? Do you see anything of significance? Make a list of the current problems.
Able to prioritise concurrent observations	What do you think is the most important problem you are faced with at present? Are you going to act in response to any of your observations? What are you going to do first?
Able to predict likely outcomes	What might happen next? Are any of the observations you have made likely to indicate that a problem may develop?
Sees conflicts between problems	How do these problems relate to one another? Are there any connections between these difficulties?
Sees options	What courses of action are open to you? What could you do?
Prioritises between options Justifies choices	What are you going to do? Why have you chosen to do that? Why did you choose not to do that? Are there circumstances in which you would have chosen a different course of action?
Adjusts to changes Explains	What will you do if..? Why does that happen? What physiology, pharmacology etc. underlies those observations and choices?
Aware of limits of judgement	Do you think you really understand that? Could you have made a mistake? Have you done the right thing? Do you now think there would have been a better choice?

current. This sort of questioning can be used to cultivate competent decision-making. The general flow of decision-making is observed without dwelling deeply on any particular issue.

Directing the trainees attention to a problem

Questions provoking thought at various levels of the cognitive taxonomy of learning objectives.[3]

Behaviour used for clinical decision making	Questions which prompt this behaviour
Perceiving with accuracy and precision	What do you see ... hear ... feel? What's happening? Tell me what happened when ...
Recognising necessary data	What facts do you need? Is that enough? What else..? How much data..? What other information? Can you decide now?
Integrating essential information	What does it mean? Put it together ... What's the problem..? What's underlying that observation?
Assessing values and determine priorities	Compare ... Contrast ... Best ... Worst ... Most important ... Least risky ... Most dangerous ... Does it matter..? Should you..?
Analysing problems situations and events	What elements..? How did you conclude..? How did you decide..? What are the possibilities?
Anticipating future events and the likelihood of their occurrence	What will happen? What might..? What would..? Under what circumstances? When? How likely..?
Making decisions based on data	You decide ... Show me ... Carry on ... You act as you think best ... Do it if or when you think its right ... On what basis..?
Projecting the outcomes of interventions	What if? If ... then what? What might..? What's the point of..? Why or when should it..?
Recognising the constraints and limitations of situations	What's stopping you? Why didn't you..? What are the limits? How far, long before you act?
Using logical inference	Why? On what basis? How did you work that out? What might be going on? Why not?
Making decisions at appropriate times	What now..? When will you..? What will you need to decide when..? Do it when you think its right.
Being efficient and organised	How did you plan for that? Could you have organised things better? What was the point of that? Could that have been done earlier, quicker?
Being flexible and responding to changes	What else? What other..? What alternatives? When will you respond to..? What made you do that?
Self evaluating one's performance and thinking	What did you think? How did you do? How good? How bad? Compare yourself? What does the surgeon feel about your performance? How could you improve? Did things go as expected?

Direct the trainee's attention to a problem and analyse it in depth. The teacher should compare the trainee's analysis of the problem with his own. He must in particular consider whether the trainee has made a reasonable assessment of the threat that the problem presents, and whether the reasons for the degree of threat and its urgency are understood.

Prompting for basic science knowledge

To improve learning of basic scientific principles the consultant must pay attention to anaesthesia as scientific process rather than a repertoire of techniques. Instruction

An example of techniques for exploring problem solving.
A 32-year-old smoker desaturates from 99% to 92% during a prone, open craniotomy for posterior fossa tumour.
Direct the trainee's thinking and problem-solving behaviours in this matter.

Has the trainee seen the problem? How timely and accurate are their observations?
- Can you see a problem?
- When did he desaturate?
- Over what period of time?
- How do you know the data is accurate?
- Did you change and check the pulse oximeter site and observe the waveform?

What does the trainee know about this problem?
- What is the significance of a saturation of 92%?
- Can you guess what the PO_2 is?
- Do we need to measure the arterial blood gases?
- How serious is a saturation of 92%? Does it need to be corrected? How soon?
- In what circumstances would an arterial blood sample be indicated?
- Does this patient have risk factors for arterial desaturation?

Can the trainee develop and evaluate a differential diagnosis for the problem?
- How do we distinguish which risk factors are important in this case?
- Can we identify a likely cause for the hypoxaemia?
- Do we need to listen to breath sounds, obtain a chest radiograph, do a fibre optic bronchoscopy?

What decisions will the trainee make?
- Should we suction the tube?
- What are the risks of coughing in this patient and what might that cause? How can we prevent that from happening?
- Should we pull the tube back blindly?
- If you diagnose a mucous plug in the right main bronchus by fibreoptic bronchoscopy, what would you do about it?
- Would you risk stimulating the carina by suctioning the plug? When?
- If so how could you prevent the adverse responses?
- What do we need to tell the surgeons?

A colleague suggests: Leave it alone and wait for closure of the dura and craniotomy. Maximise the FIO2 and ventilation. Provide a nebulised beta 1 agonist. Treat actively only if the saturation worsens. Keep the surgeons informed.
What is your response?

should concentrate on the basic principles underlying practice. Comparisons of problem-based, versus traditional undergraduate education and the aptitudes of trainees for speciality entry has suggested that an important element of success in education is the way that basic science knowledge is integrated with clinical learning. Review of underlying science, as prompted by the clinical work, is likely to be very beneficial to the trainee. The consultant should pose a problem in clinical terms and ask for an evaluation, in terms of the underlying science, or vice versa.

Using a task as a focus

The consultant can develop a special learning objective for a trainee on a list. This project will not require the trainee to be wholly responsible for anaesthesia, but to concentrate on various aspects of a problem as it occurs in different situations during the work. If this form of teaching is used, the consultant must assume responsibility for the anaesthetics and clearly identify the fact.

The task should be discussed at the start of the list, monitored throughout and be followed up at the end. A project can be used to get intensive practice or in a remedial way, when a trainee has a problem with an idea. An example would be to set the task of documenting and explaining all occasions during the list when a patient's blood pressure changes by more than 15% of the pre-op control value. When setting a task of this type, the consultant must have a clear educational purpose in mind and should make this clear from the outset. The idea is that the trainee starts by reviewing what they know, and then examines events in the light of those expectations. It may be that the trainee will need to be sent away at some stage to look things up in a textbook. The style of teaching should reflect the fact that this exercise is primarily aimed at getting the trainee to make clinical judgements.

It is also important to remember that the trainee should be able to reason out a rational course of action, and that this may not necessarily be what the consultant would do. The trainee needs to be praised and encouraged for using analytical and synthetic skills. The lesson will be lost if every exchange ends in the consultant correcting the trainee's 'mistake'. Let the decisions count unless you think they expose the patient to risk.

What difficulties can the consultant expect to find?

- The trainee will not notice important events. This will commonly be because they have not anticipated them. They will often be concentrating on something else (often the anaesthetic chart) at the time that they should be observing the patient.
- The trainee will not appreciate the importance of what they have seen. Usually this is due to lack of experience.
- The trainee will observe something, realise it is important but not know what it means.

- The trainee may have a repertoire of automatic responses that usually work and may persist with these in inappropriate circumstances.
- The trainee may not attend to the most serious problem first.

References

1. Elstein AS, Schwarz A. Clinical problem solving and diagnostic decision making: selective review of the cognitive literature. *British Medical Journal*. 2002;324:729–732.
2. Polanyi M. *The tacit dimension*. New York: Doubleday, 1967.
3. Willenkin RL. Teaching problem solving in the operating room. *Society for Education in Anaesthesia Newsletter*. Spring 1989: 2–3.
4. Epstein RM. Mindful practice. *Journal of the American Medical Association*. 1999;282: 833–839.
5. Bloom BS, Krathwohl DS. *Taxonomy of educational objectives: The classification of educational goals, by a committee of college and university examiners. Handbook I: Cognitive domain*. New York: Longmans, Green, 1956.
6. Newble D, Cannon RA. *Handbook for medical teachers*. 3rd ed. Dordrecht: Kluwer, 1994; Chapter 5.
7. Cleave-Hogg D, Benedict C. Characteristics of good anesthesia teachers. *Canadian Journal of Anaesthesia*. 1997;44:587–591.

Further reading

Abernathy CM, Hamm RM. *Surgical intuition: What it is and how to get it*. Philadelphia: Hanley & Belfus, 1995.

Brown G, Atkins M. *Effective teaching in higher education*. London: Methuen, 1988.

Bordage G. Why did I miss the diagnosis? *Academic Medicine*. 1999;74:S138–143.

Brookfield S. *Developing critical thinkers: challenging adults to explore alternative ways of thinking and acting*. Buckingham: Open University Press, 1987.

Chapman GB, Sonnenberg F, eds. *Decision making in health care: theory, psychology and applications*. New York, Cambridge: Cambridge University Press, 2000.

Elstein AS. Heuristics and biases: Selected errors in clinical reasoning. *Academic Medicine*. 1999;74:791–794.

12

Teaching practical procedures

David Greaves

There is no greater challenge for the supervising consultant than to teach practical procedures, and it is important to do this job well, as the trainee will carry the habits they learn in their first weeks of practice throughout their career.

Can skill in practical procedures be predicted?

Common sense suggests that not all trainees will be equally good at learning practical procedures. Flaaten et al. investigated the incidence of post dural puncture headache in anaesthetists of different experience.[1] Headache occurred in 56 of 394 patients anaesthetised by five trainees. There was no fall of incidence with experience, but a marked difference was seen between individuals. When Kestin investigated the learning curve of trainees undertaking a number of practical procedures the trainees showed a wide range of performance.[2]

There is no good evidence that psychomotor tests can predict future performance in tasks that demand manual dexterity.

It would be very helpful to be able to use psychometric or practical skill tests of some sort to predict who should not choose practical specialities such as anaesthesia. This is even more of a concern in surgery, where some of the practical techniques demand great manual dexterity. Shoeneman reported some correlation between neuro-psychological testing and the subsequent supervisors scores, using unstructured rating scales, for technical skills during a residency programme.[3] The best predictors were found to be complex visuo-spatial organization, stress tolerance and psychomotor abilities rather than the

Butcher. *"I often think, madam, I made a great mistake in my choice of a profession. I'm sure I should have made a fearless surgeon."*

Purdue Pegboard, a test that is considered to be a measure of manual dexterity. Another investigation examined the manual dexterity of physician and surgeon applicants to residency programmes. The Purdue Pegboard and Minnesota Manual Dexterity tests were given to 57 subjects. Analysis of the data revealed no significant difference in dexterity between medical and surgical residents, suggesting that medical students do not select specialty training programs because of the presence or absence of manual skills. The authors conclude from their data that manual dexterity tests should not be used in assessing candidates for surgical residency training positions.[4]

Gibbons reported some association of abilities to visually interpret spatial relationships (hidden figures test) and the subjective rating of surgical supervisors.[5] Reznick reviewed the evidence and concluded that it did not appear possible to predict the technical aptitude of trainees by advance testing.[6] Altmaier and colleagues investigated the relationship between critical incidents and the non-cognitive characteristics of the trainees. Over sixty percent of incidents involved non-cognitive personal attributes of preparedness, interpersonal skills and response to teaching. These authors felt that assessment of non-cognitive characteristics would improve the suitability of doctors selected for anaesthetic training.[7]

How many procedures to competence?

> *The available evidence suggests that there is a rapid improvement in the success rate with practical procedures over 20 to 30 cases but that significant improvement in success rate continues even after 80 procedures.*
>
> *No number of procedures performed will guarantee competence.*

Kestin has reported the use of the statistical method of CUSUM analysis for monitoring a trainee's progress towards mastery of a technique.[2] Though he reported only small numbers of trainees, he did document one trainee who was a very slow learner. The study demonstrated that competence could not be guaranteed, even after a very large number of attempts, and also showed that performance can get worse as well as better. It is possible that this sort of technique could be used in some practical areas to identify slow learners, in order to give them special help. Trials of this type would help identify how many repeats of a procedure are needed for most people to show acceptable competence, and answer the question whether slow learning in one practical procedure predicts slow learning in another dissimilar procedure. Kestin's data showed that there is an initial quick improvement in performance, followed by a more gradual improvement over the first 100 cases.

Konrad et al. reported the learning curve for a variety of common anaesthetic procedures.[8] They found that the curves varied between procedures with lumbar epidural being the most difficult procedure they studied. For all procedures an initial rapid improvement was seen over about 20 procedures, but significant improvements were still being seen after 70 or more attempts. Kopacz et al. studied seven residents learning epidural anaesthesia, spinal anaesthesia and endotracheal intubation.[9] They found that 20 to 25 procedures were needed in order to show improvement, but if a 90% success rate was desired 45 spinal anaesthetics and 60 epidurals were required. In an investigation by Schuepfer et al., the success rate for paediatric caudal epidural was found to be 80% after 32 procedures.[10] Martin and Burton reported the complications of one experienced, consultant ophthalmologist learning the new phacoemulsification technique. An initial vitreous loss of 4% in the first 300 cases fell to 0.7% thereafter.[11]

Marshall estimated the success of lower bowel endoscopy by measuring the learners' success in reaching the caecum.[12] First year fellows completed the examination in 54% of cases, whilst in the second year 86% of examinations were complete. The endoscopy instructors had a success rate of 97%, showing that performance can improve, even after two years practice. Over 150 cases were required for second year level efficiency, and by the end of the second year 300 endoscopies had been performed.

Rabenstein et al. investigated the impact of skill and experience of the endoscopist on the outcome of endoscopic sphincterotomy procedures. The complication rate fell with experience. They found that learning took 40 cases, but further improvement is still seen after 100 cases.[13]

Sutton et al. studied the first 150 cases by a new consultant moving towards mastery of the Ivor Lewis Oesphagectomy.[14] 150 procedures were studied as sequential cohorts of 30 cases. Improvements in morbidity, mortality, hospital stay, yield of lymph nodes and operating time were still continuing at seven years. In a similar study Parikh reported his experiences in learning a new complex procedure for gastrectomy.[15] A learning curve of about 25 procedures was determined on the basis of a morbidity scoring system. A study of the time needed to successfully complete naso-tracheal fibreoptic intubation revealed a mean time of 149 secs for the first attempt, that fell to 49 secs by the 18th repetition. The 'half life' of the curve was nine procedures. Pollock prospectively studied the complications experienced by a single neurosurgery trainee. Twenty three adverse events occurred in 728 procedures an incidence of 3%. In some categories of surgery the complication rate was as high as 9%.[16]

One group of workers has published a report suggesting a learning curve much shorter than that found by others. Yeko et al. investigated the learning of residents undertaking the laparoscopic treatment of ectopic pregnancy: The trainees had all previously performed 10 laparoscopic sterilisations, and in these circumstances the necessary learning experience was judged to be five cases.[18] Performance was judged by ability to complete the procedure and the total operating time.

Most of these reports support the view that supervision should be close for the first 20 or 30 attempts at a new procedure, but that expert performance cannot be expected until three times this number of cases has been undertaken, and that even many repetitions may not guarantee competence.

Is the learning curve associated with worse patient outcomes?

During the learning process learners without supervision have worse patient outcomes than experts. Can supervision improve these outcomes? Titley and Bracka conducted a five-year audit of training experience and the outcome of two-stage hypospadias surgery.[19] Their data suggests that whilst supervised trainees perform as well as consultants, unsupervised trainees have significantly worse outcomes. In the second stage operation the incidence of complications was 3% for consultant, 5.3% for supervised registrars but 29.6% for unsupervised registrars (fistula 15.2 stricture 4.3%). The best trainee had a complication rate of 9%, compared to 24% in the worst. The better trainees were found to have a significantly faster rate of acquiring experience. The authors suggested that having repeated experiences over a short time might help learning. Bradbury et al. were able to report the effect of trainee operators on the overall neurological complication rate in carotid endarterectomy (CEA).[20] In their hospital, between 1975 and 1991, only 12 CEAs were done by trainees. Subsequently (from 1991 to 1996) 42% of 219 CEAs were done by supervised trainees. The neurological complication rate remained 7% throughout the whole period, suggesting that the move to supervised trainee operators was as safe as the previous practice. Prasad reported that supervised trainees learning phaco-emulsification for cataracts had outcomes similar to those reported for large series with a mixture of operators. Post capsule rupture following failure of capsulorhexis fell from 20% (9 of 45) in the first 100 cases, to 2% in the next 2000 cases.[21] Kumar reported that in 433 laparoscopic cholecystectomy it was necessary to convert 14% to open operations, and the incidence was the same for consultants and supervised trainees.[22]

In the USA new interns and residents start their jobs in July. Rich, Hillson et al. examined the overall outcomes in Minneapolis and St Pauls for July, and compared teaching and non-teaching institutions. In both internal medicine and surgery, July was not associated with worse outcomes, and the length of stay of surgical patients actually declined.[23]

Most workers have shown that, with proper supervision, patient outcomes for trainees can compare favourably with those for trained specialists.

Who should teach practical procedures?

The patient is your responsibility. So ....

- Don't teach something you can't do!
- Don't teach something you have never done!
- Don't get sucked into supervising something you are not competent to retrieve.
- Never agree to let a trainee undertake a procedure with which you are unfamiliar — even if they claim to be competent in it.
- Do not direct a trainee who lacks mastery to supervise or teach another.

Trainees must be taught how to perform practical procedures correctly on the first occasion that they undertake them. They must be taught by someone who has total mastery of the technique. Indeed, a consultant who teaches practical techniques must have such mastery of them that she can anticipate, avoid and rescue every pitfall that waits for the learner. It cannot be too often stressed that no one should teach a potentially dangerous technique, until they are highly skilled in it themselves. When one trainee supervises another the teaching trainee must be fully skilled in the procedure. He or she must also have sufficient authority to be accepted as a master by the trainee being taught. I have known problems to develop when an experienced trainee has persuaded a consultant to let him do a procedure in a way unfamiliar to the consultant, or has even undertaken to teach the consultant something new. If the trainee experiences a problem they are apt, literally, to step back leaving the consultant to sort out a technique he doesn't understand.

Practical procedures must only be taught by an operator who has full mastery of the task.

It cannot be taken for granted that only a master will attempt to teach a practical procedure. I have had personal experience of trainees teaching procedures in which they are not fully proficient, and one experience of a trainee teaching a procedure they had never actually done! If a consultant orders one partially skilled trainee to teach and supervise another then it is the consultant who is guilty of incompetence.

Create sufficient time to teach

If you are going to teach a practical procedure make time for it.

- Tell the surgeon.
- Tell the other staff.
- Check that it will not disrupt the rest of the day's schedule.

Teaching practical procedures slows down the work of the surgical team. It is polite to inform surgeons, nurses and any other interested parties that there will be a delay, and it is worth organising things in such a way that delay causes the minimum inconvenience to others. This precaution also takes pressure of the trainee.

- Does the learner understand the indications for the procedure?
- Do they understand the complications of the procedure?
- Do they understand the anatomy related to the procedure?
- Do they understand how to measure the success or failure of the procedure?

Remember the patient

The welfare of the patient must always take precedence over teaching and learning. The choice of technique should always be made for the best clinical reasons and there is never a justification for adding an unnecessary procedure in order to practise it. Where patients may be injured by the activities of total novices, they should be warned of this risk and must give their consent. A frequent example of this arises when teaching the techniques of cannulation and venepuncture.

How to judge whether a procedure is a success:

- Was there a correct decision to undertake the procedure?
- Was the performance of the procedure technically satisfactory?
- Was the outcome of the procedure satisfactory?

If the patient is awake their feelings must be taken into account. It is extremely unnerving for a patient to be approached by a trainee with shaking hands and wavering needle. It is also unnerving if the consultant keeps up a commentary of possible complications. Guidance should be kept to a minimum, and must always be uncritical of the trainee and reassuring to the patient. Where things are going wrong the problems should be tactfully explained to the patient.

A system for teaching practical procedures:

1. The consultant demonstrates the procedure at full speed whilst the trainee observes.
2. The consultant demonstrates the procedure more slowly explaining each step.
3. The trainee watches the consultant do the procedure and provides an explanation of each step.
4. The trainee does the procedure with the consultant.
5. Practice – repeated performance with appropriate reduction of supervision. Mastery – will follow repeated performance and must be achieved before the trainee in turn teaches.

Preparation

If procedures are anticipated then the trainee should be asked to prepare for them. They should read the whole subject that surrounds the practical elements. They should understand the indications for a procedure and what alternatives are available. They should learn the complications of a technique and their management. They should learn about the anatomy, physiology and pharmacology involved. The anatomy is particularly important, and consultants must satisfy themselves that the trainee understands the geography of what they are to do before embarking upon it.

Unfortunately, opportunities to rehearse practical procedures may crop up unannounced. These chances should not be missed, but preparation should still take place. The teacher should take aside the trainee and explain to them the precise technicalities of the procedure, in as much as is necessary for its performance. Seeing the task demonstrated and explained is a necessary preliminary to doing the procedure. Having casually observed the technique is not enough. There is a great difference between the type and level of observation that the trainee brings to bear when they know that next time they will be asked to do the job themselves.

Position

Access is the first requirement for a successful technical performance. The patient should be carefully and optimally positioned. The operator should examine the access and carefully review whether it is adequate.

Positioning a patient on their side, or with sterile towels over their face may isolate them. It is important that an assistant, who has no role in performing the procedure, is placed in a good position to talk to the patient and hold their hand. Assistants should know how to keep up a continuous interaction with the patient, and this is a considerable skill.

Many patients will require analgesia and sedation in order to get into the position necessary for procedures to be carried out. Care must be taken with sedatives and the very minimum should be used, because this allows the operator to maintain the attention and co-operation of the patient.

Pain relief

If the patient is not anaesthetised analgesia must be effective. The absolute minimum of pain should be caused whenever undertaking a practical procedure, especially if a novice is to do the job. You will loose the patient's confidence the first time that you hurt them. When teaching, no procedure is too trivial to have adequate local anaesthesia. If a preparatory intravenous cannula is needed, this should also be pain free. I always insist on the use of local anaesthetic — a learner may need to do it twice or even three times. Will the patient remain calm if they have been hurt? I think it is a good idea to treat every adult patient as though they were a really co-operative six year old whose confidence you do not want to loose.

Let the trainee do part of the procedure – modular training

Hints for trainees:

- Know what you want to learn before you turn up for the list.
- Learn the anatomy, indications, contra-indications, complications and measures of success in advance.
- Ask to be allowed to do the procedure you want – but be prepared to compromise.
- Ask the surgeon if he or she minds you slowing the work by learning the procedure.
- If you envisage causing a delay give the surgeon and scrub team a realistic estimate of how long.
- When you are finally working alone – set yourself a time limit to accomplish the procedure. If you go over time ask for help.
- Keep a careful log of your learning.

Having prepared for the procedure by learning about it the trainee is ready to start learning how to do it.

Do not let your novice trainee begin their practical learning on a task that presents particular difficulty or danger. Find them a nice, smooth nursery slope.

The first stage of demonstration at full speed is usually superfluous, as the learner will have seen the procedure many times already. The second stage of learning, where the trainee lists the steps of the procedure, is important because the consultant can understand whether the trainee has got all the steps in mind, and in the right sequence — without needing to ask.

When stage three is reached, and the trainee is the operator, the consultant must remain calm and guide the trainee with as little interference as possible. Distracting the trainee is not helpful. It is essential that the learner commentates on their own performance as they go along. This allows the consultant to watch without interruption, because the trainee has made clear the matters that they have in mind. The consultant does not need to ask what is coming next, because the trainee is going to tell them.

Repetition is essential for learning, and particularly when practical skills are being learned. When a procedure is complex, it is difficult to provide meaningful repetition, because the variety of different tasks confuses the learner. This problem can be helped by letting the novice do part of the procedure each time. This way, they get a number of opportunities to repeat the same discrete section of the procedure. They then practice different individual sections, and finally undertake the whole task.

A further argument in favour of modular training for practical procedures is that the learner can become exhausted by the work, both mentally and physically. I have known a consultant to make his trainee do several successive spinal anaesthetics for a urology list. The learner was not good at the procedure, and became distressed because of the delay they were causing, and exhausted by the mental effort of coping with a series of difficult procedures. The consultant thought they were giving the trainee a really good learning opportunity.

I adopt a modular approach when teaching epidural anaesthesia. I break the procedure into four sections.

- Assembling and testing the apparatus.
- Location of the space, skin infiltration and puncture through to the ligamentum flavum.
- Location of the space.
- Threading the cannula, securing the appliance and giving a test-dose.

The same approach can be taken when setting up a major case. If the trainee performs the insertion of an arterial line, central venous line and thoracic epidural, a slight delay is expected. Allowing learners to do procedures causes great delay to work, and it is sensible to adopt this modular approach in order to keep things moving. Causing delay on an operating list is actually dangerous. The surgical team become pressured, and in turn they put pressure on the anaesthetic team to hurry through the rest of the work. I have personally had experience of a surgeon embarking on the wrong procedure, because they were hurrying to catch up after they had let their trainee take a long time over a simple procedure.

Slowness with practical procedures is not necessarily a sign of incompetence, but expert operators are always steadily moving forwards with their work. When the operator does not know what to do and stops being purposeful, then the case progresses slowly. Excessive slowness is a reason for investigating outcome.

When they are learning, trainees are often at the limit of what they can cope with. In the jargon they are dealing with a high task load. It is the consultant's responsibility to lighten the load to what they can cope with. This means not challenging them with too much, or something too difficult. It also means helping, by taking some of the peripheral load. The consultant should monitor the patient, adjust fluids, control the anaesthetic and deal with the surgical and nursing teams. This leaves the trainee free to concentrate on the procedure.

There are no hard and fast rules as to what can be permitted, and it is an exercise of the consultant's professionalism to set the balance between training, safety and expedition. Never let a trainee persist, unsuccessfully with a technique. In the interests of both the learner and the patient, the consultant should be quick to step in and get the job done well. This is in the trainee's best interest, as the chances of a good outcome with a procedure are less as time passes, and it is better to take over early than to let failure drag on. Bystanders and impatient surgeons must be kept well out of the way.

A problem for consultants is the part-trained learner who is using a technique that is not part of your usual repertoire. Experienced teachers should attempt to broaden their practice so that they can cope with common variations, and departments should attempt to standardise practice between those consultants that teach. If the technique that the trainee wishes to use is outside your experience, then you must consider whether they will be able to complete their learning with the technique they have started. If there are few opportunities to work with those who are familiar with that variation, then you should start them off again with your local technique. Do not supervise something you are unfamiliar with. If a senior trainee is genuinely master of a technique — let him/her do it. If you are convinced that the trainee is good enough to teach others including you, accept this gracefully.

Do not distract the trainee

The supervisor must not distract the trainee from the task in hand. All direction must be helpful, and it is particularly important not to overload them by asking difficult questions when their minds are fully occupied with the task in hand. The consultant must not seem to become visibly alarmed whilst the trainee is performing a procedure. If the trainee sees that concern, he or she will loose confidence and may become panicky. Cool assured supervision is the order of the day. Even experienced doctors are often anxious about practical situations, but they must learn to stay cool.

Review

Every practical teaching session should be followed by a review. Feedback, as always, should be positive. The trainee should be praised for the things that went well, and positive advice given for the things that went badly. The important thing is to make sure that the trainee knows what went badly and accepts this. Sometimes they have no insight and feel that, even though the outcome was not successful, they are now competent to go on to practice alone.

At the end of each feedback the trainee should be told the consultants assessment of the overall performance, and what level of supervision they will need next time they attempt this procedure.

The importance of continuing help and support beyond the initial learning stage was demonstrated in a paper that reviewed the complication rate of surgeons participating in a course to learn laparascopic cholecystectomy. Surgeons who proceeded to work alone after the course were three times more likely to have a major complication than those who continued to seek help and supervision, or attended courses to learn more.[24] A particularly revealing study was that of Davies and Campbell, who examined the experience of trainee surgeons at the time that they undertook their first 'solo' inguinal hernia repair. Trainees had undertaken between two and ninety supervised repairs (median 8).[25] Seventy nine percent of the group said that they felt competent to perform the procedure alone at the time of their first 'solo'. I do not find it reassuring that twenty percent of trainees were prepared, unsupervised, to embark upon a procedure in which they did not feel competent.

References

1. Flaaten H, Berg CM, Brekke S, Holmaas G, Natvik C, Varughese K. Effect of experience with spinal anaesthesia on the development of post-dural puncture complications. *Acta Anaesthesiologica Scandinavica*. 1999;43:37–44.
2. Kestin IG. A statistical approach to measuring the competence of anaesthetic trainees at practical procedures. *British Journal of Anaesthesiology*. 1995;75:805–809.
3. Schueneman AL, Pickleman J, Hesslein R, Freeark RJ. Neuropsychologic predictors of operative skill among general surgery residents. *Surgery*. 1984;96:288–295.
4. Squire D, Giachino AA, Profitt AW, Heaney C. Objective comparison of manual dexterity in physicians and surgeons. *Canadian Journal of Surgery*. 1989;32:467–470.
5. Gibbons RD, Baker RJ, Skinner DB. Field articulation testing: a predictor of technical skills in surgical residents. *Journal of Surgical Research*. 1986;41:53–57.
6. Reznick RK et al. In: Bartlett RH, Strodel WE, Zelenock GB, Harper ML, Turcotte JG, eds. *Medical education — a surgical perspective*. Chelsea MI: Lewis Publishers Inc, 1986.
7. Altmaier EM, From RP, Pearson KS, Gorbatenko-Roth KG, Ugolini KA. A prospective study to select and evaluate anesthesiology residents: phase I, the critical incident technique. *Journal of Clinical Anesthesia*. 1997;9:629–636.
8. Konrad C, Schupfer G, Wietlisbach M, Gerber H. Learning manual skills in anesthesiology: Is there a recommended number of cases for anesthetic procedures? *Anesthesia and Analgesia*. 1998;86:635–639.
9. Kopacz D, Neal J, Pollock J. The regional anesthesia "learning curve": what is the minimum number of epidural and spinal blocks to reach consistency? *Regional Anesthesia*. 1996; 21:182–190.

10. Schuepfer G, Konrad C, Schmeck J, Poortmans G, Staffelbach B, Johr M. Generating a learning curve for pediatric caudal epidural blocks: an empirical evaluation of technical skills in novice and experienced anesthetists. *Regional Anesthesia and Pain Medicine*. 2000;25:385–388.
11. Martin KR, Burton RL. The phacoemulsification learning curve: per-operative complications in the first 3000 cases of an experienced surgeon. *Eye*. 2000;14:190–195.
12. Marshall JB. Technical proficiency of trainees performing colonoscopy: a learning curve. *Gastrointestinal Endoscopy*. 1995;42:371–373.
13. Rabenstein T, Schneider HT, Nicklas M, Ruppert T, Katalinic A, Hahn EG, Ell C. Impact of skill and experience of the endoscopist on the outcome of endoscopic sphincterotomy techniques. *Gastrointestinal Endoscopy*. 1999;50:628–636.
14. Sutton DN, Wayman J, Griffin SM. Learning curve in oesophageal cancer surgery. *British Journal of Surgery*. 1998;85:1309–1402.
15. Parikh D, Johnson M, Chagla L, Lowe D, McCullogh P. D2 gastrectomy: lessons from a prospective audit of the learning curve. *British Journal of Surgery*. 1996;83: 1595–1599.
16. Pollock JR, Hayward RD. Adverse operative events in neurosurgical training: incidence, trends and proposals for prevention. *British Journal of Neurosurgery*. 2001;15:312–318.
17. Smith JE, Jackson AP, Hurdley J, Clifton PJ. Learning curves for fibreoptic naso-tracheal intubation. *Anaesthesia*. 1997;52:101–106.
18. Yeko TR, Villa A, Parsons AK, Maroulis GB. Laparoscopic treatment of ectopic pregnancy: Residents' learning experience. *Journal of Reproductive Medicine*. 1994;39: 854–856.
19. Titley OG, Bracka A. A 5-year audit of trainees experience and outcomes with two-stage hypospadias surgery. *British Journal of Plastic Surgery*. 1998;51:370–375.
20. Bradbury AW, Brittenden J, Murie JA, Jenkins AM, Ruckley CV. Supervised training in carotid endarterectomy is safe. *British Journal of Surgery*. 1997;84:1708–1710.
21. Prasad S. Phacoemulsification learning curve. *Journal of Cataract and Refractive Surgery*. 1998;24:73–77.
22. Kumar A, Thombare MM, Sikora SS, Saxena R, Kapoor VK, Kaushik SP. Morbidity and mortality of laparoscopic cholecystectomy in an institutional setup. *Journal of Laparoendoscopic Surgery*. 1996;6:393–397.
23. Rich EC, Hillson SD, Dowd B, Morris N. Speciality differences in the "July phenomenen" for twin cities teaching hospitals. *Medical Care*. 1993;31:73–83.
24. Cooper WA. Predictors of laparoscopic complications after formal training in laparoscopic surgery. *Journal of the American Medical Association*. 1993;270:2689–2692.
25. Davies BW, Campbell WB. Inguinal hernia repair: see one, do one, teach one. *Annals of the Royal College of Surgeons England*. 1995;77(suppl):229–301.

13

Informal mini-tutorials in the operating theatre

David Greaves

Surgery is often protracted, and the anaesthesia may proceed without significant event for a number of hours. Such periods when consultant and trainee are together, are opportunities for informal mini-tutorials. Indeed, these are what many consultants and trainees understand by the term theatre teaching. Trainees may deny that there has been any teaching on a list, unless a mini-tutorial has taken place! Two studies have found that anaesthetists believe that the ability to relate practice to basic science theory is a marker of excellence.[1,2] Theatre teaching can help put basic science into a clinical context, and this aspect of learning should be a major focus for the in-theatre mini-tutorial.

Teaching topics should use the context of the anaesthetic — this draws attention to the patient rather than creating a distraction.

Mini-tutorials are just very small group teaching events and they can be treated as lectures, tutorials, problem based learning sessions or any other type of teaching that appeals to the consultant.

Is it safe to teach mini-topics in the operating theatre?

Teaching about theory in the operating theatre is not dangerous — as long as it is conducted properly.

There has always been controversy about whether it is safe to teach anaesthetic topics in the operating theatre. There is not a lot of evidence one way or the other. Some years ago Boulton suggested that quiet times in theatre make ideal opportunities for teaching, but was unable to provide any evidence to support his view that it was safe.[3] In another review of theatre teaching it was held that teaching except in relation to immediate events could not be defended.[4] There has been some suggestion that critical incidents have been reported as stemming from teaching activities, but there have not been any full evaluations of these incidents to analyse whether the teaching style was responsible. Data from 2000 critical incidents reported to the Australian Incident Monitoring Survey did not reveal any relationship between distraction due to teaching and critical incidents, though inattention (a skill failing) was present as a factor in 10% of incidents.[5] Cooper et al. found that teaching was in progress and was an associated factor in the genesis of 60 out of 507 critical incidents.[6]

Lambert and Paget demonstrated in an analysis of theatre behaviour that teaching did not appear to distract consultants from their patients.[7] No consideration was paid to the type of teaching in progress.

Mini-tutorials are probably no more distracting than all the other activities that the anaesthetic team may engage in during long surgery. It is easy for attention to wander and anaesthetists who are sitting watching the monitors carefully may be inattentive. Advice on a safe framework for teaching mini tutorials is offered throughout this chapter. Situated learning by mini-tutorials is, in this authors opinion, an essential part of teaching and learning in anaesthesia.

Planning mini-topics

Use mini-tutorials to help the trainee to think about problem solving.

Mini-topics should be short. Ten minutes is enough. More than fifteen minutes is too much. They should also be limited to one or two a day. More than this turns the theatre into a classroom and everyone gets bored. Mini-topic teaching is a distraction that is particularly likely to annoy the

A context-based scenario for discussion of the physics of anaesthetic circuits.

Context:

Any spontaneously breathing patient on a Bain circuit.

Scenario:

What would happen to the patients end tidal CO_2 if the flow rate of fresh were reduced to equal minute ventilation. Why did Maples classify anaesthetic circuits in the way he did?

surgeon, and many consultants are quite insensitive in this regard, allowing the audience for their talk to grow until they are addressing the whole theatre and can probably be heard in the coffee lounge.

Topic teaching is an important way of getting a trainee to relate events in the operating theatre to basic science, and for getting him to use his faculties for critical evaluation. Discussion and topic teaching should be reserved for times during the operation when the patient is physiologically stable. When a consultant chooses to teach a trainee around a topic, a situation analogous to bedside teaching in the ward, he must firstly take over care of the patient and formally identify this fact to the trainee:

"I am going to talk to you about the control of blood pressure and whilst I am doing this I will take full responsibility for the care of the patient."

1. Choose a topic relevant to the patient.
2. Review the patient and take responsibility for monitoring and record keeping.
3. Present a scenario to the trainee.
4. If appropriate send them away to think about the problem.
5. Draw together the conclusions and sum them up in the context of the present patient.
6. Consider suggesting follow-up tasks.

This both sets the trainee free to listen, and indicates that he should attend to the proposed teaching. It is common to see a consultant talking to a trainee in theatre, who is trying to listen, whilst giving half his attention to the patient. That trainee is not going to remember anything. The teaching consultant must be seen to be carefully observing the patient's progress. If the trainee feels that the patient is being neglected he will become distracted. It is a surprise to consultants to discover that trainees do not always entirely trust them as practical anaesthetists. A trainee can be a severe taskmaster, and they cannot be taught if they are not satisfied with the conduct of the anaesthetic. Experienced consultants may note a problem and discount its relevance without comment. The trainee then becomes distracted by what seems to be failure of observation, and is not in the right frame of mind to learn. Indeed, they will apply themselves to vigilantly monitoring the anaesthetic, to compensate for the consultants apparent lack of attention.

Occasionally, consultants try to teach at times when there is too much happening in the theatre for either them, or their trainee, to give their minds to education. Occasionally, a consultant is seen harassing a trainee with questions, whilst the trainee has his hands full keeping the anaesthetic under control. If it becomes clear that the trainee is unable to concentrate on the teaching, then it should be abandoned. Some consultants cannot teach, and some trainees cannot learn in theatre in this semi-didactic way.

A context-based scenario for a short discussion of the regulation of blood pressure.

Context:

A hip revision is being performed. An epidural anaesthetic is being supplemented with a propofol infusion.

Scenario:

Let us imagine that this patient's blood pressure fell to 60/30. I want you to think about the different changes in cardiovascular physiology we might see if this blood pressure drop were due to haemorrhage, extensive epidural block or acute myocardial infarction.

A theatre topic discussion should have relevance to the case in hand. Not only does this help to keep attention on the patient, but it also improves the contextual relevance of the teaching. There is some evidence that teaching is better remembered if it was taught in the sort of circumstances in which it will need to be recalled. There have been reports of school children that can remember learned material much better in the classroom, or their bedroom, than they can in an examination hall. Educational psychologists believe that the memory of information is accompanied inevitably by memory of the context and situation in which it is learned. This is called the context dependency effect, and it was demonstrated by Godden and Baddeley in a classic experiment.[8] They set two groups of divers a task learning lists of words. One group learned on the beach, whilst the other learned under fifteen feet of water in full SCUBA. The beach learning group subsequently remembered the material better on the beach whereas the SCUBA learners had better recall under water. Recall was 40% better when testing took place in the same medium as learning. A similar result was obtained by Smith, Glenber and Bjork.[9] Their subjects learned lists of words in a room with an experimenter, and were then tested either in the same room with the same experimenter present, or with a different experimenter in a different room. Recall was superior when the conditions for learning and testing were the same. Interestingly, recall for material learned when drunk or under the influence of drugs is superior when again drunk or spaced out. The tentative conclusion can be drawn that book-based revision in the trainees flat may be less effective in providing them with timely remembrance of important facts, than rehearsal of the same facts in the acute operating theatre, where they will be needed. Unfortunately we can also conclude that the best learning for an MCQ test taking place in an examination hall, may be different from the best learning for the real workplace.

Topics can be related to practical anaesthesia, but the theatre situation can also be used to show the relevance of aspects of physiology and pharmacology. This is a

A context-based scenario for a short discussion of distribution of agents injected intra-venously.

Context:

An elderly patient is being anaesthetised for a routine laparotomy. Induction was with thiopentone.

Scenario:

What do you think would have happened if we had used a dose of 8 mg/kg of thiopentone to induce anaesthesia in this patient? What would have happened if we had given that dose to a fit young man of 23? How can the differences be explained?

particularly important lesson, as trainees often fail to carry over their understanding of basic sciences into clinical situations.

The teaching can be more active if the trainee is given some time to organise and muster their thoughts. One technique is to present the trainee with a clinical scenario or a sample of data around which the teaching will centre. They are then given ten minutes to go away and think. On their return the consultant takes them through the relevant issues. Context based scenarios should be composed by looking at the anaesthetic in progress, and deciding upon a principle that is currently being applied. A scenario is then composed, which dwells upon the basic medical science surrounding that principle. Looking at the appropriate taxonomic level for such episodes of teaching, the nature of the context and the presence of an expert moderator suggests that the levels of synthesis and evaluation are most appropriate. Scenarios that call for appraisal, comparison, criticism, and discrimination, whilst asking for supporting evidence, will prompt learning at that level. Discussion should then centre on the justification for the conclusions that have been made. The learner should be asked to speculate about relevant outcomes and then to defend his reasoning by reference back to basic principles. Many trainees will answer questions of the 'what if' variety by producing a list of possibilities. They must then be made to justify and criticise their listing.

The content of a mini-topic can be used repeatedly during the remainder of the operating list.

It is worthwhile to prepare teaching materials for context based mini-topic teaching in the operating theatre. The operating theatre allows vivid demonstration of many of the principles of physiology and pharmacology. Unfortunately the underlying basic science is not always easy for the consultant to remember, and this frequently discourages them from

Prepare yourself some cards listing the 'hard to remember' facts that will enable you to revue a topic well.

embarking upon teaching. More senior trainees who appear to be very knowledgeable often intimidate consultants. Part of the consultant's repertoire of teaching scripts can be materials for teaching in this situation. Each scenario relates to relevant theatre contexts in which they should be used. A scenario or discussion topic is provided, along with a few short answer or multiple-choice questions. The material also includes any 'hard to remember' data that the consultant will need to provide an explanation of the answers to the questions. Many times I have staggered to a halt in full flow of deriving the Henderson–Hasselbach equation, explaining the alveolar gas equation or drawing the internals of Dr Von Recklinghausen's oscillo-tonometer. A relatively small prompt was all that was needed for me to be able to complete a good presentation. That is the purpose of the mini-topic teaching material.

A variation on topic-based teaching is Service-based learning (SBL). This is one of a number of educational programmes that are based on the principles of Experiential Learning. It is noteworthy as an attempt to incorporate learning into the daily work schedules of trainees, and as having been developed upon a careful analysis of the way more junior trainees work.[10] It is not a new way of teaching or learning but an alternative way of structuring learning in order to derive maximum benefit from clinical experiences.

There are four strands to the SBL:

- Practice-based learning.
- Use of brief learning materials and self assessment questions.
- Educational meetings.
- Study time.

SBL was developed in response to the observation that in the United Kingdom SHOs spend a large amount of their time working independently in providing service. As a consequence of the emphasis on service provision, the quality of the teaching and learning that SHOs receive has been very variable.

Practice-based learning

The theme of this strand of SBL is to turn clinical experience, which might be viewed as purely service provision, into systematic education.[11] This shifts the emphasis and the perception of the trainee and the trainer to realise that much of their service work is actually an opportunity for learning.

There are a number of key elements, which are necessary to ensure that the trainee approaches their education in a systematic way. There must be a core curriculum, so that the trainee knows what is expected of them. They need an educational supervisor to guide them. They need to learn how to reflect on their clinical practice, so that they can identify those aspects that are satisfactory, and those aspects that need attention.

As a consequence of this approach each trainee should develop an individual learning plan or contract in conjunction with their educational supervisor. The plan should identify what knowledge, skills and attitudes the trainee needs to learn. It should identify how the learning will be achieved, by what date, and how success will be assessed. Implicit in this plan is the requirement for regular reviews or appraisals (formative assessment). These enable the educational supervisor monitor the trainee's progress, so that modifications can be made as necessary.

This approach to using a trainee's experiences as the basis for learning is not new. I am sure many of us adopted this approach during our own training, but perhaps in a less formalised way. I certainly remember the more effective and popular teachers in my training always seemed to be better organised, and knew the direction in which to guide their trainees.

Brief-learning materials

What happens if a trainee identifies, from their clinical experience, a gap in their knowledge, skills, or attitudes? If they are in the company of a senior anaesthetist they might ask them. If they are lucky they will get a short, concise and structured answer, combined with an opportunity for discussion. They should then either have resolved the area of ignorance, or planned how they are going to build on this experience. If there is no senior available, or they are reluctant to discuss their ignorance, where should they go? Ideally, there should be learning resources readily available. These resources should be understandable and relevant, along the lines of a quality discussion with a senior anaesthetist. Academic textbooks are rarely a suitable resource, as they have usually been written with different aims in mind. These resources also need to be readily available, not locked up in a library that is often a long way from the clinical areas.

The proposal for SBL is to develop Brief-learning materials (BLM) for the 'core curriculum'. A BLM can take any format but each should take less than 15 minutes to complete. As part of the BLM there is a self-assessment with the ideal answers.

It is envisaged that the trainees might use the BLMs in three separate ways:

- When they need instant information for their daily work.
- When a learning need arises from their experience-based learning and is identified in their educational contract.
- When they are working through the curriculum ensuring they have no gaps in their knowledge or experience.

Meetings (Formal training)

This strand consists of and builds on the present structured classroom teaching that is already in place.

There is the formal teaching that is also referred to as 'protected teaching' time. This strand tends to address the knowledge or theoretical aspect of the content of the curriculum. It can also be used to aid trainees' critical thinking and presentation skills. It also gives the opportunity for trainees at the same stage in their training to get together to share ideas.

There are the departmental meetings, which will include audit, mortality and morbidity, journal clubs and clinically orientated subjects.

Part of this strand is the crammer course, designed to prepare them to take and pass the examinations that are now part of the requirement to proceed to the next stage of their training.

Study time

The final strand is enabling the trainees to plan their study time. This consists of helping and encouraging them to utilise their time away from the clinical areas effectively.

The intention of SBL is to help the trainees integrate the academic side of their learning with their clinical experience. Sometimes trainees, and trainers for that matter, view the academic or theoretical side of their learning as only relevant to passing the formal examinations. As it is easier and more effective for adults to learn when they see it is as relevant to themselves, the art is to integrate the two aspects.

SBL has been successfully implemented in a number of hospitals for Anaesthesia, Accident and Emergency Medicine and Paediatrics.[12] These programmes are available from the Joint Centre for Education in Medicine.

A lot of attention has focussed on the use and design of brief learning materials. The important element of SBL is, however, the way in which it makes an educational plan that attempts to use service, classrooms and books to meet well-defined and individualised objectives. It has also been unusual in that it looked at the actual working patterns of clinicians, and attempted to fit teaching around these, rather than packaging education in ways that did not fit well with working practices.

References

1. Cleave-Hogg D, Benedict C. Characteristics of good anesthesia teachers. *Canadian Journal of Anaesthesia*. 1997;44:587–591.
2. Klemola U-M, Norros L. Analysis of the clinical behaviour of anaesthetists: recognition of uncertainty as a basis for practice. *Medical Education*. 1997;31:449–456.
3. Boulton TB. Teaching in the operating theatre (Editorial). *Anaesthesia*. 1980;35(5): 451–453.
4. Tammisto T. Teaching in the operating room. *Cahiers d'Anaesthesiologie*. 1988;36(6): 479–494.
5. Cooper JB, Newbower RS, Kitz RJ. An analysis of major errors and equipment failures in anaesthesia management: considerations for prevention and detection. *Anaesthesiology*. 1984;60:34–42.

6. Runciman WB, Sellen A, Webb RK et al. Errors, incidents and accidents in anaesthetic practice. *Anaesthesia and Intensive Care*. 1993;21:506–519.
7. Lambert TF, Paget NS. Teaching and learning in the operating theatre. *Anaesthesia and Intensive Care*. 1976;4:301–303.
8. Godden DR, Baddeley AD. Context dependent memory in two natural environments: On land and under water. *British Journal of Psychology*. 1975;66:325–331.
9. Smith SM, Glenber A, Bjork RA. Environmental context and human memory. *Memory and Cognition*. 1978;6:342–353.
10. Grant J, Marsden P. *Training senior house officers by service based learning*. London: Joint Centre for in Medicine Education, 1992.
11. Stanton F, Grant J. Approaches to experiential learning, course delivery and validation in medicine. A background document. *Medical Education*. 1999;33:282–297.
12. *Service Based Learning: Anaesthesia*. London: Joint Centre for Education in Medicine, 1996.

14

Teaching anaesthesia to new starters

David Greaves

It is important that novices are shown good practice from the start of their careers. If trainees are not taught the correct way to do things they will develop bad habits that it is difficult to get them to change later in their career. This chapter is in many ways the most controversial of this book because it represents a very personal view. Some will disagree fundamentally with some of the advice offered, and undoubtedly there are good alternative approaches to the problems discussed here. In defence of what I have written I can only offer that it comes from nearly three decades of teaching novice anaesthetists, observing their difficulties and searching for ways to minimise the many pitfalls that await them as they move to independent practice.

> *Teach the trainees always to ask themselves whether each action they contemplate taking will improve matters.*

Preliminary reading

Trainees are usually keen to get started and they should be provided with some hints for reading when they are appointed. Every anaesthetic department should produce some sort of departmental handbook and this should also be given to the trainee. The handbook should include useful information such as a plan of the hospital and information about the operating theatres. It should include details of the anaesthetic department staff and information on how the department is organised. It should also include copies of clinical policies and protocols that have been adopted by the department. The handbook given to new starters should include a special section for them. This should describe how they will be taught anaesthesia and indicate the level of supervision they should expect. Where training is to follow a curriculum, a copy of this should be given to the trainee. Some anaesthetic departments have produced a booklet that describes simple anaesthetic techniques and includes lessons for the first week of anaesthesia.

It should, however, be assumed that the new trainee will not have read the materials provided! The first days of anaesthetic practice should be conducted on the assumption that they know nothing. This avoids misunderstandings.

There are two schools of thought about how to begin teaching anaesthetic trainees. Some teachers advocate starting off with a block of theoretical instruction in the classroom. This will cover the important practical and theoretical knowledge that the trainee needs to have in mind as he gives his first anaesthetics. The second approach, which I favour, is to plunge the trainee straight into theatre on day one. My reason for preferring this approach is to make maximal use of the enthusiasm that the trainee brings to his new trade, and to take advantage of the vivid experience of theatre for experiential learning. Anaesthetic practice is so different to the medicine that most trainees have been engaged in that they do not comprehend the context of the theory one tries to teach them, until they have become familiar with the operating theatre. In Newcastle upon Tyne we allow new trainees four weeks experience before we take them out of theatre for an intensive course of theory and practice.

The induction day

Make sure that the anaesthetic department gets plenty of time with the new trainees on induction day.

How to fill in an accident form is important — but....

Hospitals will have induction days for new doctors. These will be multidisciplinary, but often there is time for departments to take part of the day. If the formal induction day is filled with talks about pharmacy, radiology forms and so on it is essential that the anaesthetic department make time to present its own departmental introductions.

Establishing a framework for new starters

Anaesthetic practice is very confusing to new learners. The organisation of the first weeks of training should take account of this, and be directed towards creating as much stability as possible. Especially in big hospitals, new trainees should work in a very restricted number of operating theatres. Experienced theatre workers are inclined to forget how alien the environment was when they first started in it. Nowadays neither medical students nor house officers spend much time in theatre and new anaesthetists will probably not know how to behave. They should be properly taught how to behave in theatre, and it is better if a consultant does this in the course of work, during the trainee's first visit to theatre. This is much less embarrassing to the trainee than being paraded around theatre in a group by one of the theatre sisters. The trainee should be told about local policy on issues such as wearing masks. They should be told where they can walk in theatre and which doors they can go in and out of. It is very distressing for a trainee to be shouted at for using the wrong door or walking under the ventilation hood, when no one has told them otherwise. Trainees should be introduced to staff by

> Allocate new starters to appropriate simple short anaesthetic procedures.
>
> Allocate new-starters to a few consultants who specialise in their supervision.
>
> Teach new-starters a restricted range of techniques.

name, and this should include everyone they come across including the surgeons. Coffee arrangements should be explained, including local taboos on which coffee rooms are within bounds and any arrangements for paying for drinks in which the trainee will be expected to participate.

Why have I dwelt at such length on the social aspects of the operating theatre? It is essential that a new trainee is happy and enjoying work right from the start. What we want to do is encourage enthusiasm and this is all too easily dampened by a few unpleasant experiences, caused by not knowing what to do.

One specific area that should be dealt with, as soon as trainees start to work in theatre, is that of sterile procedures and precautions. New trainees often do not understand the elementary principles of theatre sterile technique, such as how to open packs and sterile packets. They should be taught properly how to scrub and put on gloves and gowns even if they claim to be expert already.

New trainees should work only in theatres that have a suitable type of surgery. This will include general surgery, urology, orthopaedics and gynaecology. Within those areas they should be allocated to a suitable mix of cases. Working repeatedly in a theatre where one long case is being done restricts their experience of starting and stopping anaesthesia, and working with short day cases denies them any experience of monitoring patients for longer cases.

Trainees should be taught a small range of techniques. This is best achieved by restricting the number of consultant teachers with whom they work and by agreeing a range of practices that all these teachers will use. Within these selected techniques there should then be some standardisation of approach.

Receiving new trainees into the department is quite a complicated exercise and will benefit from some planning. Looking after the new trainees should be made the responsibility of one consultant. A meeting should be held before the date when the new starters arrive, and plans made for their reception. Appropriate lists should be identified in sufficient numbers to accommodate the number of new starters. A number of consultant teachers should be identified, and the range of anaesthetic techniques to be used should be decided. Finally, a working allocation of the trainees should be drawn up. Some departments do not make allocations for new trainees, and prefer to let them move around to where they can find appropriate work. Many trainees are very diffident about doing this, and some are frightened. I believe it is preferable to make firm allocation of trainees so that they know where to go and what they are doing. If the work in their allocated area is not suitable, then their consultant can redirect them. He should take them to the new workplace, introduce them to the new supervisor and explain what they should be doing. It goes without saying that new starters should not be allocated to consultants who are on holiday!

In the United States where all new residents start in July it is customary to have a 'one to one' period of two months where trainees are assigned to specific attendings (consultants) for their training. Each attending works with a resident for one or two weeks. Concurrently the resident will attend an introductory lecture series. During this period the resident is assessed and less intense supervision (one attending to two residents) will be allowed for work where the resident is sufficiently competent. One to one training will then be re-instated when the trainee moves to a new sub-specialty.

A theory course for new starters

Right from the start trainees need to know some rudiments of theory. A programme of discussions or mini-tutorials can be used to cover critical issues. This can be formal classroom teaching but, as the tutorials may only be suitable for one or two trainees, many departments prefer to give the trainee a list of topics that are checked off as individual consultants cover them during the course of work. Teaching this early basis for the theory of practice in the context of the real work helps the trainee's understanding.

Practical skills

There should be a list of practical skills that the trainee must master before undertaking independent practice. In the UK some of these are now included in the mandatory first SHO assessment that is included in the Royal College of Anaesthetist's Guide to Training.[1]

Planning supervision

Novice anaesthetists should be strictly supervised. The Royal College of anaesthetists has identified three levels of supervision.[2]

New starter anaesthetists should be supervised at level 1 for the first three months of their experience. They should not proceed to level 3 until they have six months

The Royal College of Anaesthetists' definitions of levels of supervision:

- Immediately available supervision means the supervisor is actually with the trainee or can be called within seconds.
- Local supervision means that the supervisor is on the same geographical site, is immediately available for advice and is able to be with the trainee in ten minutes of being called.
- Distant supervision means the supervisor is rapidly available for advice but is off the hospital site.

experience. Experience does not necessarily imply competence and teachers must find some formal way of deciding that a trainee is competent enough to move to a lower level of supervision, or take on a new responsibility. This can be done simply by discussion amongst the consultants. It is sensible to make such discussions formal, and to make a record of decisions. An alternative is to use a competence assessment to decide whether the trainee should move on. If this is the departmental policy, then the test should be applied formally and a record kept. The public, and managers on their behalf, require that trainee doctors are adequately supervised, and in cases of mishap the doctors who manage training will need to be able to show that their processes are not haphazard and that the trainees have been assessed before being moved on. Self evidently, trainees must be suitable for work with level 3 supervision before they take any responsibility for out of hours work.

The first day in theatre

Beginners want, and need, *simple rules that they can follow.*

Trainee anaesthetists can start doing things as soon as they start in theatre. The trainee needs to understand the separate components of the anaesthetic, but always to see them as a coherent whole. There should be some preliminary instruction by the consultant. This should take place away from the patient, so that hearing their anaesthetist being given remarkably basic instructions does not alarm them. Trainees must be reminded that they will need to talk to the patient as they prepare for anaesthesia, and must make clear explanations of what they are doing. They must be reminded that the patient does not understand many of the words they take for granted. It is common to hear theatre personnel chatting to the patient about 'monitoring' and 'electrodes', quite unaware that the patient has no idea what they are talking about.

A good starting point is showing them how to draw up the drugs for a case. The trainee can then go on to putting in a cannula. Part of this early period of training should be to review the skills of cannulation. Despite their previous experience many trainees have never been shown how to insert a cannula correctly, and their technique tends to be a little unpredictable. As the patient goes off to sleep the trainee should

An anaesthetic SHO with six months experience anaesthetised a ten-year-old boy in the Accident and Emergency Department. He administered thiopentone. The child developed catastrophic bronchospasm and died before a consultant could reach the scene (it was night-time and the consultant was at home).

At the coroners inquest the SHO was criticised for not knowing that thiopentone was probably contraindicated in asthmatics. His consultants were criticised for not having taught him this fact.

This case occurred over ten years ago and clearly the exact situation should not be repeated now. The principle remains that even a conscientious team of consultants could miss out some fundamental piece of teaching without which their trainees are in danger of harming patients.

be instructed in the maintenance of the airway. The best sort of anaesthetic to start a trainee with is that of spontaneous breathing with a facemask and airway. Control of the airway with bag and mask is the fundamental technique of airway management that every anaesthetist must have immediately at their command. These days many of these short cases would be managed with laryngeal mask airways, but wherever possible new trainees should be asked to maintain the airway with a bag and mask, until the technique is second nature to them.

There is a tendency to break the anaesthetic sequence down into sections when dealing with novice trainees, and many consultants like to move their trainee around the theatre to get lots of experience of induction and reversal. There is a contrary view that trainees should always be encouraged to think about anaesthesia as a whole from start to finish. My preference is very much for this latter approach. I insist that the trainee see the patient before the operation, administers the anaesthetic and then follows them through to recovery. It is useful for trainees to recover the patients themselves, so that they quickly learn about the problems of waking up. In this way the trainee is taught to consider the patient rather than the anaesthetic. Consultants should also encourage trainees to complete all the paperwork relating to cases. Right from the outset they should be taught to think about postoperative pain and nausea. They should prescribe the postoperative analgesia, antiemetics and fluids. Every case should be entered in the trainee's new logbook. There is no substitute for the satisfaction the trainee feels having given whole anaesthetics on their first day. In my experience it is sensible to let the trainee play a major role in alternate anaesthetics. This gives them opportunity to follow their patient to recovery, and to meet their next patient at the theatre reception.

Developing skills

New anaesthetic skills should always be added on the foundation of what has been previously learned. They should be encouraged to discuss the patients' experiences

The three stages of early training:

1. The first few weeks: Struggling to make rules. Nothing makes sense. They are oblivious of most problems.
2. The next few weeks: Skilled in simple practical procedures. Improving but paradoxically feel they are worse because they now see all the things that go wrong around them.
3. After 2 to 6 months: Advanced beginner. Now seeking for exceptions to the rules.

with them, and in particular to give attention to the frequent problems of post-operative nausea and vomiting.

As each new technique is learned it is important that the teacher emphasises the reasons for choosing a different way of doing things, and the ways that it differs from the options that the trainee has already encountered. The consultant should link the new to the old by discussing the alternatives.

It is not advisable for trainees to learn to conduct anaesthesia according to a recipe book of techniques. Trainees who have just begun anaesthetics will automatically try to learn a repertoire of ways of doing things that they will then use in response to a list of indications that they may also learn by rote. Whilst this may not be a problem in the early weeks of practice, they will eventually have problems, unless they can use sound reasons for choosing between options.

Trainees should also be taught to see the consequences of what they choose to do, and should be taught to proceed only when they are confident of recovering from any likely complications.

New starter anaesthetists go through three stages in their very early training. For the first few weeks they are completely at sea. Nothing makes any sense to them, and they have to be told how to do everything. During this stage of training they are searching for rules and consistent ways of practising. Many supervising specialists are tempted to teach them that anaesthesia is chaos and that nothing can be relied upon. This is not what they need to hear. At this stage, they need help to perceive the basic consistencies of practice. This first stage of training usually lasts about six weeks.

The next thing that occurs, as the trainee improves, is that they think they are getting worse. The reason for this is that they are now able to see the things that are going wrong around them. In the early weeks of practice their anaesthetics are often out of control, and the consultant is quietly engaged in keeping things on an even keel. The trainee is focussed on a relatively few activities, such as airway control, and does not see what goes on. As they become more technically proficient, they begin to see all the problems that surround them and they think that they are getting worse. At this stage they need to be reassured, and told why things seem worse. If they are not reassured their fragile practice deteriorates, because their decision-making is apt to freeze. This second stage of practice lasts anything from two to six months.

Finally the trainee reaches the stage of being an advanced beginner. At this stage they have reasonable situation awareness in straightforward cases. They have a framework of rules from which they work, and now they are starting to learn the exceptions to their rules. This is a critical stage of training, because they must be guided in the direction of working out a plan, when an exception occurs, rather than finding another rule. If they continue to build rules they will soon be overwhelmed.

How long does all this take? It is immensely variable. It is important not to compare the progress of trainees with that of their peers until they have done twelve months. I have known many trainees who were still in the second stage of their progress at six months, and about whom colleagues were expressing concern, who have gone on to be excellent consultants. Slow progress through stage two is not an indication that progress will always be slow. If the exigencies of on-call lead the consultants to press the trainee who is progressing slowly they can damage their confidence so badly that they will leave anaesthesia.

> *Having done a number of cases does not guarantee competence*
>
> *— but you can't be competent if you haven't had this minimum experience.*

The problem of anaesthesia becoming light

A great many problems that are encountered by novice anaesthetists arise from the airway problems that are caused by anaesthesia becoming too light. This situation is exacerbated by the fact that the trainee usually believes that the patient is too deep! Very serious difficulties can be precipitated, such as bronchospasm and regurgitation and aspiration of gastric contents. So common are these difficulties that it is worth explaining to trainees how they often arise from lack of attention to detail in giving anaesthesia. Trainees often seem unable to understand the importance of obtaining smooth transition from intra-venous induction agents to maintenance, inhalational anaesthesia.

All these difficulties result in the patient becoming light, and coughing and struggling. Unfortunately this is particularly likely to happen during transfer of the patient to theatre. Trainees should be advised always to take their time over establishing maintenance anaesthesia, before moving the patient or allowing the surgeon to start.

The problem of the apnoeic patient

Another common difficulty for trainees is re-establishing spontaneous ventilation following intra-venous induction of anaesthesia. In modern practice induction with propofol, particularly when synergised by opiates, often results in apnoea. Many anaesthetists count this as a bonus, because they are then able to hand ventilate the patient and enhance the uptake of inhaled anaesthetic. For novices, however, it causes problems,

Problems that arise for novice trainees if the patient stops breathing at induction:

- They may not be able to ventilate the patient with bag and mask.
- Their efforts to ventilate a light patient who is breath holding may cause the patient to cough.
- They may inflate the stomach with gas.
- The patient may become light as the intra-venous agent redistributes without any inhalational agent being breathed in.

both in managing the airway and in re-establishing breathing. Sometimes a trainee will over-ventilate the patient and use high concentrations of inhaled agent. This depresses the patient's respiratory drive, and half an hour later they may still be trying to get the patient to breathe. Alternatively, because they are aware of this problem, they reduce the anaesthesia and the patient becomes too light. This latter problem is very common for novices who are using the laryngeal mask airway in a patient that received a potent opiate with induction. My response to this problem, when using propofol for induction, is to point out to trainees that the agent crosses the blood brain barrier slowly, and that it should be given at the slow rate of 40 mg in 10 seconds as described in the data sheet. At this rate propofol does not cause significant respiratory depression, and the patient should continue to breathe spontaneously and allow plenty of time for the inhaled agent to rise to adequate alveolar concentration.

Consultants will render patients apnoeic deliberately when inducing with propofol because it allows them to ventilate the patient with the inhalational anaesthetic agent....

... This is a high level skill and is often beyond the abilities of a beginner.

The laryngeal mask

Though the laryngeal mask airway is a boon to anaesthetists, it causes many problems for trainees. Novices have difficulty deciding when to use the device and must be given clear guidelines. Major problems can arise from inappropriate applications, and it is important that trainees understand where there are areas of controversy. Many consultants will make use of the laryngeal mask for gynaecological laparoscopy. Setting aside any general objections to this practice, it is not safe for inexperienced trainees. Consultants should not depart from mainstream practice at all when teaching very raw trainees, and must always tell a trainee of any grade when they are departing from standard practice. Trainees must be alert to the difficulty of placing laryngeal masks correctly, and must understand that it will still work as an airway, even when seriously misplaced. I have found that airway difficulties resulting from misplaced laryngeal masks are very common during the first year of anaesthetic training.

It is the small points of difference in practice that confuse trainees. I have focussed on some simple problems with anaesthesia, not only because they are the cause of a large number of the problems that trainees meet during their first few weeks, but also because they illustrate the level of agreement over technique that a team of teaching consultants should achieve. Variety of approach is very beneficial to training, but not at the outset. Consultants responsible for initial training should agree a policy for all these issues. When there is a cause for confusion between them the trainees will usually ask about it. The policy should then be extended to accommodate this additional point. Anaesthetists are notoriously dogmatic in the defence of the minor idiosyncrasies of technique that they espouse, and new trainees are a captive audience for an airing of these views. Education is much better served if these minor differences can be shelved whilst new trainees are being taught.

Resuscitation training

New trainees should have formal sessions on advanced life support as soon as they start anaesthesia. No doctor wearing a badge identifying them as a trainee anaesthetist should be subjected to the indignity of not knowing what to do at a cardiac arrest. Though the novice should not have any duties that require them to be part of a resuscitation team they may, in the course of their duties, be first on the scene.

Use of a topic logbook

An example of how a topic could be described.

How intra-venous induction agents are distributed.

Outcome: The trainee has a basic understanding of the principles of redistribution and elimination of intra-venous induction agents sufficient to allow him/her to:

- Choose appropriate doses.
- Give injections at appropriate rates.
- Adjust doses for sick patients.

Guidance:

Clinical context: Induction of patients using a variety of induction agents including Propofol TIVA.

Theoretical issues: Three phases of distribution, effect of cardiac output, rate of crossing the blood brain barrier, different types of half lives.

New trainees need to be instructed on thirty or forty separate topics during their first few weeks of anaesthetic practice. One way of dealing with this is to organise a course of tutorials. This takes the trainee away from the operating theatre. A better plan is to make use of a logbook for discussion topics. A list of the key topics for discussion is provided to each trainee, and at appropriate times during the day in the operating theatre the consultant talks about one of the topics on the list.

- Use of a list ensures that the full introductory curriculum is covered.
- Teaching is in context and therefore likely to be remembered.
- Teaching associated with experience is best.
- Efficient use is made of both the consultant and the trainees time.

Providing a list of topics however does not necessarily provide optimum teaching for the new starter.

In the USA where most residency programs have a mass influx of residents in July, tutorials covering the basics of anesthesia are arranged over this initial period. The use of topic checklists for attendings is also frequent.

Should there be a list of essential knowledge within each topic?

Perhaps it might be a good idea to subdivide some topics to give the consultants some idea of the scope and depth that you are expecting. A full guide to the consultant would state the objectives of the lesson, the outcome in the workplace, the appropriate time to choose to demonstrate this lesson and the list of theoretical issues to be covered. Each topic should be ticked off on a list that the trainee keeps with him or her. The record should be kept for annual review.

Could you structure the experience in order to facilitate learning?

If you find that there is a clear order of learning which works well you may want to consider how best to provide the contexts that promote teaching of these topics. Perhaps these contexts arise regularly in routine work. Sometimes, it will be sensible to make sure that a trainee is despatched to an appropriate list, so that they can receive teaching about something at the right stage of their clinical development, and in context. An example might be to send a trainee to the obstetric unit, to learn about the complications of local anaesthetic toxicity and their treatment. It is sometimes very helpful for consultants to teach a topic and then send the trainee off to see other kinds of work that reinforce the points that have been made.

References

1. *The CCST in anaesthesia II: Competency based senior house officer training and assessment, a manual for trainees and trainers.* London: The Royal College of Anaesthetists, 2000.
2. *The CCST in anaesthesia I: General principles, a manual for trainees and trainers.* London: The Royal College of Anaesthetists, 2000.

15

Teaching anaesthetists how to behave in an acceptable professional manner

David Greaves

Relationships in the workplace

An effective clinician possesses:

- Knowledge
- Skill
- A repertoire of behaviours which permit him or her to deliver patient care

and always delivers these in a professional manner.

It has been my experience, from long involvement with anaesthetic trainees, and their problems, that dysfunctional behaviour, though infrequent, is as least as common as examination failure as a cause of career difficulty. In the course of consultant practice, behavioural problems lead to years of conflict and under-performance, that sometimes culminate in suspension and dismissal. The stresses between surgeons, anaesthetists and nurses working in the operating theatre are the stuff of popular drama, but in real life they often complicate the work and cause unhappiness.

"Relax. He's in a good mood."

An anaesthetist attends an emergency in the accident and emergency department and correctly diagnoses a tension pneumothorax. The rest of the team, which includes other equally senior doctors from other disciplines, is already there and have misinterpreted the clinical signs and are treating the patient for acute left ventricular failure. The correct outcome in this situation is for the team to change direction and follow the anaesthetist's lead towards a successful conclusion. Whether this occurs depends very much upon the behaviour of the various doctors, and this sort of situation all too readily leads to a stand-off with catastrophic consequences to the patient. It is often forgotten that part of learning is getting to know how to behave. Where a task requires complex interpersonal relationships and depends on teamwork, learned behaviour is vitally important. It is as important to be able to lead a team and inspire them with the confidence that you know what you are doing, as it is to know what to do in the first place.

To be effective a doctor must be able:

- To make his clinical thoughts understood to the team.
- To pay attention to the clinical thoughts of other team members.
- When appropriate, to lead the theatre team.
- When appropriate, to follow the lead of another member of the team.
- To deal with his or her emotions, including anger and fear, so that they do not interfere with work.
- To deal with the emotions, including anger and fear, of other team members so that they do not interfere with work.

It has already been stressed in earlier chapters that the operating theatre is a good place to learn how to behave.

What skills does the doctor need to learn?

- Team working
- Team leadership
- How to follow a leader

Particular problems occur when the roles in the team have to shift. It is easy for conflict to arise. The most damaging pathologies of teamwork are a primary failure of leadership and conflict leading to a paralysis of leadership.

Relationships with patients

Doctors are under pressure from the public, their patients and politicians. Partly this reflects the failure of doctors to move with the times. Paternalistic attitudes and an impenetrable aura of arcane knowledge were, in earlier times, not merely the normal demeanour of doctors, but to some extent the public expected and was reassured by them. An apocryphal story that probably has a basis in fact recounts how a professor of surgery had made the incision for an inguinal herniorrhaphy before he realised that the patient was presenting for haemorrhoids. It was subsequently explained to the patient that the roots of the piles had been so deep that they had required excision from above. This tale is not entirely incredible to anaesthetists of my generation. In our post-modern world experts of all kinds are not trusted. Technical knowledge no longer gives a remit for unquestioned authority. Interestingly, there is a tendency to attribute the changed milieu to financial improprieties in the USA and to high profile criminal and civil cases in the UK. Probably the real reasons are similar on both sides of the Atlantic, and simply reflect changes in the attitude of the public to the professions in general.

Whatever the causes for mistrust, the public has spoken clearly about what it expects of the medical profession and we must respond. Changing behaviour is the basis of education, so the proper behaviour of doctors has become an important concern of medical teachers.

What does the public expect of doctors?

In the United Kingdom the General Medical Council has issued guidance for doctors that describes the 'Duties of the Doctor' in detail.

The American Board of Internal Medicine has developed a project to define and analyse professionalism in medicine.

- A commitment to the highest standards of excellence in the practice of medicine and in the generation and dissemination of knowledge.
- A commitment to sustain the interests and welfare of patients.
- A commitment to be responsive to the health needs of society.

The Duties of the Doctor.[2]

Patients must be able to trust doctors with their lives and well-being. To justify that trust, we as a profession have a duty to maintain a good standard of practice and care and to show respect for human life. In particular as a doctor you must:

- Make the care of the patient your first concern.
- Treat every patient politely and considerately.
- Respect patients' dignity and privacy.
- Listen to patients and respect their views.
- Give patients information in a way that they can understand.
- Respect the right of patients to be fully involved in decisions about their care.
- Keep your knowledge and skills up to date.
- Recognise the limits of your professional competence.
- Be honest and trustworthy.
- Respect and protect confidential information.
- Make sure your personal beliefs do not prejudice your patient's care.
- Act quickly to protect patients from risk if you have good reason to believe that you or a colleague may not be fit to practice.
- Avoid abusing your position as a doctor.
- Work with colleagues in the ways that best serve patients' interests.

In all these matters you must never discriminate unfairly against your patients or colleagues. And you must be prepared to justify your actions to them.

The qualities of practice identified were:

- Altruism
- Accountability
- Excellence
- Duty
- Honour and Integrity
- Respect for others

Seven issues that challenge or diminish professionalism were identified.

- Abuse of power
- Arrogance
- Greed
- Misrepresentation
- Impairment
- Lack of conscientiousness
- Conflicts of interest

The American Academy on Physician and Patient has examined the doctor patient relationship from a similar perspective. They have concentrated particularly on aspects of communication.

- Relationship building
  - Listen for feelings and respond actively.
- Information gathering
  - Elicit patients biomedical and psychosocial data efficiently.
- Patient education
  - Tell a diagnosis meaningfully.
  - Negotiate and implement a plan.
- Non-verbal skills
  - Enhance alliance, safety, trust and bi-directional information transfer.

The Royal College of Anaesthetists and the Association of Anaesthetists have in the UK issued similar guidance for individual anaesthetists and departments.

What is the importance of all this information and advice for anaesthetic teachers and learners? All these documents represent attempts to describe the proper and moral behaviour of doctors. Most doctors have always acted in accordance with these tenets. But there are those who abuse their position, by behaving in ways that range from the mildly unethical to the grossly criminal. One particular failure that the public perceives is that doctors cover up their own deficient and dangerous practice and tolerate that of their colleagues.

How do we learn how to behave?

Good role models are the key to teaching professional behaviour. When a trainee works with a consultant they see how to behave. Consultants behave in many ways and there is a role model that suits the character traits of every trainee. This is probably what determines the adoption of role models rather than any conscious choice. Ideally a trainee should watch the consultant and analyse his or her effectiveness and the impact of behaviour on outcome. They should then aspire to what is good, and reject what is bad. Unfortunately the process is much less cerebral than this. The trainee watches the consultant and may admire their effectiveness. Effective behaviour may include bullying, aggression, clinical deceit and greed. Though these do not serve the best interest of the team, or the patient, they may be very successful in satisfying the requirements of the doctor. They are seen to embody the essence of the successful anaesthetic consultant. The trainee then begins to behave in the same way. Behaviour that works for an established consultant may not work for a new SHO and they may rapidly run into difficulty.

1. *Tell them how anaesthetic professionals should behave.*
2. *Set them an example. Learn how to be a good role model.*
3. *Adopt good role models.*

Professionalism:

- A commitment to the highest standards of excellence in the practice of medicine and in the generation and dissemination of knowledge.
- A commitment to sustain the interest and welfare of patients.
- A commitment to be responsive to the health needs of society.

How to teach and learn good behaviour

The consultant must understand that, whenever they teach a trainee, there is a hidden lesson available to be learned from the consultant's behaviour. The lesson may be quite simple — this is how big shot consultant anaesthetists behave. From this, the trainee learns how to be rude to surgeons and nurses, how to bully and bluster, how to obstruct and create petty difficulties for the work. If there are few alternative role models, then this is how they will learn to behave. There may well be some truth in stereotypes of difficult professionals, such as neuro and cardiac surgeons! An alternative lesson is learned from the weak ineffectual anaesthetist who gives in to the surgeon, even if this is not best for the patient and who allows the team to be dominated and made miserable. Other role models are as aggressive as Jack Russell Terriers and about as predictable, or lazy and ceaselessly complaining.

The consultants must realise the importance of their behaviour. It doesn't matter what you tell the trainee to do, they will probably choose to do what you do. If you are rude, you spawn a new generation of rude anaesthetists. It is probably easier to adopt bad role models than good, because many of the features of good behaviour are necessarily unobtrusive. Quiet, firm control of a situation may not be noticed; blazing conflict will be, and if the outcome is successful for *our* side then the lesson will be learned!

Consultants should openly discuss theatre behaviour. Not only will this help the trainee to choose good behaviours, but it is also good therapy for the consultant. It is an opportunity to reflect upon one's own conduct and consider whether it bears inspection. Consultants should talk about their personal strategies for exerting leadership, learning to listen to others, giving and receiving orders and dealing with rudeness, aggression and anger in others. There are no right answers to any of these issues but it is often useful to explore alternatives.

Post-operative debriefing should include comment on how the trainee has dealt with relationships and his or her role in the team. This should also be included in appraisal discussions and assessment should take these matters into consideration.

How to learn how to behave

Trainees should give some thought to the impact of behaviour in the workplace. The most important thing to realise is that good behaviour simplifies life, gets the work done and avoids conflict. Above all it improves patient management. Always be on the lookout for difficulty developing and try and work out why it is happening. There is no doubt that the behaviour of many doctors is outrageous. Try always to remain calm and present your point of view in a rational manner. Never ever respond to irrational behaviour in kind. Never use patients as bargaining points or blackmail. Never ever argue with a colleague in the acute situation. Find some sort of temporary resolution, and then discuss matters with the person concerned away from the theatre. Remember how it will look if catastrophe strikes and the anaesthetist and surgeons have been having a public row. If things are going badly wrong, get in touch with your consultant.

Be aware of changes in attitude about medical care and engage in the debates. When I was a junior trainee patients were rarely told the truth about terminal illness, and senior consultants would lie to patients about their prognosis and about specific elements of their diagnosis and treatments. Dishonesty was common when clinical errors had occurred. These things were done with the best possible motives. Times have changed, and now most clinicians are frank with their patients. This change came about through a new generation of doctors adopting a different set of beliefs about their patients' rights, despite the continuing alternative role model of their more senior colleagues.

Organise your professional life. You cannot do two things at once. Do not create problems by being unpunctual. Always give yourself time to prepare properly. Don't put yourself in the wrong by arriving late. Many problems arise because the anaesthetist arrives late and is hurried into action before he or she has got things sorted out properly. Always introduce yourself to people you work with. Never work with a surgeon until you have introduced yourself. At emergency situations always announce yourself by name, grade and specialty.

Don't get overtired and hungry. Many trainee anaesthetists have been worn down by being expected to continue working with relays of different surgeons. Always take a meal break and insist upon a rest from time to time.

Dealing with problem trainees in the operating theatre

Trainees often develop minor difficulties with their work and learning in the operating theatre.

At least one in ten trainees is, at some time during their training, perceived to have a problem in the way they undertake their work, or in the way in which they are progressing in training. These problems can appear because the trainee is unskilful, ignorant or does not behave acceptably. Such problems can be very serious and may indicate illness or serious social difficulties. Everyone that deals with trainees must look out for the signs of serious difficulties and these must be dealt with by professionals. Many trainees, however, have minor problems that appear in their theatre work and which can easily be sorted out by their supervising consultants. It is counter-productive to refer such trainees to occupational health for counselling.

Lack of skill

Genuinely ham-fisted trainees are uncommon. Perhaps trainees who lack manual dexterity do not attempt a career in such a technical specialty. Compared to surgery, the practical procedures of anaesthesia are few and they are mostly short and with well defined end points of success or failure.

Minor clumsiness is unfortunately common. Post-anaesthetic morbidity such as haematoma and lip trauma is reported in up to 30% of patients. It is often necessary to draw this matter to a trainee's attention. A suitable challenge is to ask them to do a list of short minor cases without causing any injuries at all. This is surprisingly difficult to achieve. Most of these complications can be avoided just by taking care.

It does not appear possible to predict problems due to lack of manual dexterity, but it is probably possible to detect such problems by auditing success and failure at specific procedures. Problems with a technical procedure can result from faulty technique, which is commonly due to the trainee not having been taught sound technique when they first undertook the procedure. If a trainee is known to have a problem with a technique it is appropriate to re-teach the technique as though they had not done it before. They will often then realise where their technique varies from the ideal. An alternative is to watch the trainee do a procedure, but in my experience it is often difficult to see where the problem lies.

Lack of knowledge

The operating theatre is not a good environment for teaching knowledge, but lack of relevant understanding will quickly reveal itself during clinical work. Practical anaesthesia has to be founded on proper knowledge and it is common to see trainees acting as though they did not understand what they are doing.

Teaching can address this issue. The consultant should work with the trainee and, from time to time, ask them to explain the relevant pharmacology and physiology of what they are doing. Where necessary he or she should give short explanations. Where the trainee's working knowledge is felt to be unacceptable, they should be told to go and read about it. The instruction should tell them exactly what to read, and should be followed up by a question and answer session. Usually this scheme is effective.

Dealing with trainees with behavioural difficulties

These sorts of problems are relatively common. Difficulties encountered range from conflict due to inappropriate aggression and anger, to absenteeism due to extreme fear and stress. Where difficulties are major, then the trainee will need full diagnosis and counselling. This discussion is limited to a consideration of minor difficulties which affect performance in theatre and which a consultant can help during normal supervision and teaching.

A senior trainee regularly left the operating theatre to go to the theatre kitchen and sitting room. He did not respond to direct instruction not to do it. His consultants co-operated in making him observe a strict routine of feeling the patient's pulse and taking the blood pressure manually. After some weeks of this regime he appeared to have got into the habit of observing the patient closely and thereafter his behaviour was acceptable.

A trainee was arriving late in theatre regularly. This was causing difficulty both in starting emergencies and in running routine lists. She accepted that she did this and had no excuse except her general disorganisation. In discussion it was clear that she had not really considered the effect of her behaviour on her colleagues, neither had she realised that this in turn led them to be unfriendly and aggressive. She made some efforts to keep better time.

A trainee was becoming aggressive and causing problems for the whole theatre team towards the end of the working day. He agreed that he knew his behaviour was difficult and explained that he was constantly worried about the possibility of the list running late which would interfere with him collecting his child from the crèche. A change to his child minding arrangements coupled with an accommodation by the rota-maker resolved his difficulties and the bad behaviour disappeared.

A trainee often became personally abusive during stressful clinical situations. Discussion showed only that the problem was complex and irrational. Referral to the occupational health physician and thence to a psychiatrist lead eventually to an acrimonious decision to quit anaesthesia.

A trainee was very unsure of her clinical judgement and often felt she was being criticised by colleagues including quite junior nurses. Even necessary requests for instruction were met with aggression. The doctor readily accepted that there was a problem and initially appeared to accept reassurance that her clinical judgement was excellent. The problems continued until she was brought to understand that her behaviour was the very reason for the guarded suspicion with which she found herself regarded by colleagues.

References

1. *Maintaining good medical practice*. General Medical Council, 1998.
2. *Good medical practice*. General Medical Council, 1999.
3. American Board of Internal Medicine. *Project professionalism 1995*. Available from: http://www.abim.org
4. American Academy on Physician and Patient. 1998. Available from: http://www.physicianpatient.org

16

Making routine judgements about clinical competence

David Greaves

Performance is what the doctor *does*, and from repeated measures of performance it is possible to make a judgement of competence — what the doctor *can* do.

Though assessment in the operating theatre is usually informal it has significant consequence for the trainee. If such judgements are incorporated into any review of the trainee's progress, then they must be fair, and fairness demands that the processes of making a judgement are as careful as for any other test or exam. They must also be open. There is no place for allowing consultants to make anonymous complaints about trainees.

Consultants need to understand the clinical competence of trainees:

- To plan their progress and to decide what to teach.
- To match supervision to the trainees level of skill.
- To provide evidence of clinical progress to any formal review of progress.

At present most assessment in the operating theatre is not structured. The trainee has not been provided with instructions about what they are to learn, and to what standard they must perform. The consultant does not discuss this with the trainee and may not even know exactly how experienced the trainee is The trainee is not usually told how they are being judged. Often, there is no direct feedback to the trainee from the consultant. The trainee does not even know that a judgement has been made, let alone what it was. It cannot be too strongly emphasised that any judgement that feeds into the processes that determine the progress of trainees must be done in accordance with good practice. If in-service assessment is part of the formal system for the assessment of specialists for accreditation, then it must be able to stand legal challenge.

Two definitions of competence:

- Competence in a job is defined as the ability to perform the tasks and roles required to the expected standard.
- Competence is adequate clinical proficiency as judged against the performance of an accepted specialist.

How can we measure clinical competence in the operating theatre?

In-training assessment should only be introduced with careful preparation. The stages for the introduction of an in-training assessment are:

- *Define what is to be tested.* The assessment must be directed towards the required aspects of performance. It is essential that proper objectives are defined in a curriculum and that these are known to the trainees and the assessors.
- *Establish machinery for testing.* If adequate preparation is not made then assessment will generate a lot of results that cannot be correlated, and do not contribute to the education process. Specialists will have to be taught how to conduct the assessments. Secretarial resources will be needed and mechanisms will have to be organised for following up and counselling trainees. It will also be necessary to define what action will be taken in the case of trainees who are judged incompetent.
- *Select a test to use.* Decide upon a type of test to apply. Organise it and apply it.
- *Agree standards for the assessment.* Look at the results from pilot assessments and use these to develop standards that must be reasonable and consistent. There are a number of strategies for the direct in-service assessment of proficiency. They are all dependent on the judgements of supervising specialists, in one way or another, and despite concerns that observation of practice is a very subjective assessment such judgements have been found to be useful.

There are a number of ways to incorporate the observations consultants make of trainees at work into an assessment system.

Daily In-Training Assessment (ITA)

In-training assessment is systematic observation and feedback on the habitual performance of trainees. ITA provides information in many areas of a trainee's performance that are difficult to assess by other means, including clinical skills, clinical judgement, communication, interpersonal behaviour and attitude to patients. All the trainers make observations, and they are able to assess progress as the trainees pass through training.

> *Critical comments about trainees made during in-service assessment must always include specific examples with details.*

Consultants who work with trainees are asked to return a daily report by completing a form. Questions relate to a number of categories of performance that usually include skill, knowledge and measures of attitude and professionalism.[1, 2] Scoring is usually on a rating-scale with additional room for spontaneous comments. Enthusiasts contend that this daily review system is very efficient at identifying trainees who are having difficulty.[3] Using many observations on many occasions improves the content-validity and overall reliability of the assessment. Daily reports on performance must be very easy to complete or compliance will be poor. The data may therefore be somewhat shallow.

There has been very little attention paid to how in-training evaluation reports correlate with other measures of clinical performance. A Japanese group has used a rating scale for the daily assessment of trainees.[4] Both the trainee and supervisor scored their performance against a number of practical criteria (e.g. stability of blood pressure). They found the scoring to be consistent and the trainees scored themselves an average one point lower than the supervisors.

ITA is used for annual review of trainees in many training programmes and unsatisfactory reports will ultimately lead to termination of training. Parenti and Harris conducted an analysis of how faculty formed opinion on the performance of house-staff in everyday practice.[5] They found that a group of experienced faculty members believed that there were aspects of performance that discriminated well between

Categories for in-service assessment

POSTGRADUATE INSTITUTE FOR MEDICINE & DENTISTRY & THE NORTHERN SCHOOLS OF ANAESTHESIA
COLLEGE TUTORS SUMMATION OF ASSESSMENTS

Gradings

	Appropriate to experience	Cause for concern	Unacceptable	Unable to comment
Clinical skills				
Knowledge				
Practical skills				
Personal characteristics				
Initiative				
Judgement				
Manner				
Organisational ability				
Communication skills				
Time keeping & reliability				
Departmental involvement				

A behavioural rating scale for trainee assessment with three points on the scale. Anchors describing acceptable and unacceptable behaviour are shown on the following page.

Behavioural anchors for in-training assessment

	Appropriate to grade	Cause for concern	Unacceptable
Clinical skills			
Knowledge	Adequate and up to date.	Occasional gaps in knowledge.	Lacks essentials. Poor ability to apply knowledge.
Judgement	Normally good application of knowledge. Appropriately seeks advice.	Poor application of knowledge. May fail to ask for help when necessary.	Unreliable. Fails to grasp significance of situations. Fails to recognise limitations and seek advice appropriately.
Practical skills	Normally good.	Difficulty with some procedures.	Poor skills for stage of training.
Personal characteristics			
Initiative	Normally shows initiative, takes responsibility appropriately.	Needs pushing and may fail to show initiative. Slow to take responsibility.	No initiative. Does not take responsibility.
Manner	Good sense of team. Good working relationships.	May be careless of others. May have difficulty team working. May make rather than solve problems.	May be rude or arrogant. Careless of others. Poor sense of team. Causes rather than solves problems.
Organisational ability	Normally well prepared and organised. Deals competently with admin tasks. Adapts to local policies.	May be unprepared and poorly organised. Muddles some admin tasks. Slow to adapt to local policies.	Poorly prepared and disorganised. Unreliable with admin tasks. Fails to adapt to local policies.
Communication skills	Good communicator. Establishes rapport. Listens well.	Sometimes has communication difficulties with staff, patients or relatives.	Often has communication difficulties with staff, patients or relatives.
Time keeping & reliability	Punctual and reliable. Warns department of problems.	Sometimes late or unreliable. Has failed to warn of problems.	Often late or unreliable. Usually does not warn of problems.
Involvement in department	Participates in departmental activities.	Participation below that expected.	Rarely participates in any departmental activity.
Confidence			
	Over confident	**Appropriately confident**	**Under confident**
Confidence	Works beyond level of knowledge, training or experience.	Assesses situations accurately. Undertakes tasks or asks for help appropriately.	Inappropriately hesitant, worried or over cautious for grade.

This is a limen referenced scale for competencies depending on the consultant's judgements.

Cardiopulmonary resuscitation

The Trainee:

This assessment may be undertaken at any time during the first month of anaesthesia. Assessment may be combined with a *practical* teaching session.

	Yes	No
Understands sequences for single handed and assisted basic CPR.	☐	☐
Satisfactorily interprets common dysrhythmias on ECG monitor.	☐	☐
Demonstrates satisfactory cardiac compression.	☐	☐
Demonstrates good mouth to mouth rescue breathing.	☐	☐
Can undertake the lead role in directing CPR.	☐	☐
Demonstrates ventilation with bag and mask.	☐	☐
Understands the indications for defibrillation.	☐	☐
Demonstrates correct use of defibrillator.	☐	☐
Understands the appropriate use of drugs during resuscitation.	☐	☐
Understands the dangers of CPR to the resuscitation team.	☐	☐
This assessment was completed satisfactorily.	☐	☐

individuals. They suggested that such aspects of performance should be given greater importance in overall assessment schedules.

One study found positive correlation between consultants' scoring for non-cognitive skills (conscientiousness, confidence etc.) and the incidence of clinical near-miss incidents.[6] ITA is used for annual review of trainees in many training programmes, and unsatisfactory reports may lead to termination of training. Where an adverse comment about performance is made it should always be backed up by a specific example, and should include a note of the date and time and any witnesses.

Periodic reporting of performance

The format is similar to the daily evaluation method, except that supervisors are asked to provide a summary of their overall impressions at the end of a period of training. A report form is used, usually with scores allocated to specific categories of behaviour. The consultant reports are then sometimes discussed within the department and a summary produced.[7] This form of assessment is essentially similar to daily reporting, but can be criticised as it relies more on memory, often weeks or months after the events. Fewer observations are taken, which may make the scoring less valid. On the other hand, as assessment is less frequent, consultants may be willing to put more effort into completing the forms.

Practice observations

A practice observation takes place where a specialist works with a trainee and specifically scores and records his or her performance during the course of work. The observation takes place on an operating list that has been chosen as being appropriate to the competencies being assessed, and is effectively a type of examination. Observations can

When making assessments in the workplace observations should:

- Take place in a familiar setting.
- Involve real work.
- The trainee must know that assessment is in progress.
- The trainee must know what aspects of performance are being observed.
- The trainee must know what level of performance is required at the level of the assessment.
- The trainee should understand any scoring criteria used.
- The outcome should be explained to the trainee who should be able to comment in some detail.

be of work in general, or they can be targeted on a particular activity such as communication skills or vigilance. This sort of approach has not been used very much for the evaluation of anaesthetists. This approach produces a score that tries to reflect how the trainee has performed, rather than concentrating on what has been done. It differs from the previous assessment system in being more like an examination. The consultant is specifically observing a repertoire of theatre behaviours and scoring them directly. It is likely that consultants will need training in order to implement this type of assessment.

The format of assessment:

- Tell the trainee what you are going to expect them to achieve in order to be judged satisfactory.
- Tell them how you will make the judgement of whether they are satisfactory.
- Make an assessment.
- Explain to the trainee what you have decided and upon what evidence you have based your opinion.

Practice observation must be as objective as possible. A checklist of observations should be used, which includes careful guidance to observers. The supervisor watches the trainee and uses the list to check that all key features of management are undertaken. Each item can also be scored against a scale of performance, in which case clear guidance should be given on the scoring. Examples of good and bad practice are noted by the observer and reviewed to provide an overall scoring of performance.

It is important to understand that this approach to scoring does not address the fundamental question of standard setting. Consultants who believe they are using the same criteria may be producing very different reports.

The outcome of an unsuccessful practice observation should not just be reporting a fail. Feedback should be given, and targets should be set prior to undertaking further observation. Records should be kept and carefully correlated, so that any trends can be seen.

Criterion-referenced competencies

In order to make an in theatre assessment of a trainee:

- The consultant should have worked with the trainee often enough to have observed a reasonable sample of the trainee's behaviour.
- The consultant must know the previous experience of the trainee.
- Any summative assessment should represent the pooled judgements of several observers on a number of occasions.
- The consultants must know what skills abilities are required of the trainee as set down in the curriculum.

Competency based training is a fashionable form of education that has been used in vocational training and it particularly applies to teaching skills for the workplace. It has been discussed in Chapter 3. A criterion-referenced competency is one that has an outcome that can be measured against a standard. This is often a standard derived from the consultant's clinical judgement as to what is good practice, but more specific criteria are practicable and useful when assessing practical procedures and aspects of knowledge.

In such a scheme the list of necessary competences takes the place of a traditional curriculum. As a competence is acquired the learners behaviour changes, and this can be observed. This is the outcome of the competence. It is listed along with the competence and assessment consists of observing the outcomes. A competency based approach to learning must therefore list all the competencies needed along with the outcomes, and guidance as to how the outcomes should be investigated.

The positive outcomes of direct assessment of clinical competence.

For the learner:
- It judges mastery of essential knowledge and skills. It measures progress with time.
- It can detect individual areas of weakness.
- It sets the agenda for learning professional behaviour.

For the process of learning:
- It reinforces the extent of the syllabus.
- It defines the rate of learning.
- It increases emphasis on professional behaviour as a learning activity.

For the teaching:
- Success tends to predict future success.
- It improves theatre supervision and teaching.
- It fosters situated learning.
- It helps discover why some things are not learned.

For clinical practice:
- It safeguards the interests of the public by setting standards of practice in the workplace.
- It provides a basis for specialist certification.

A competence based approach to teaching and assessing practical anaesthesia would involve a number of steps:

- Analyse the tasks of anaesthesia and define the curriculum in terms of observable outcomes or competencies. e.g. 'correctly performs venous cannulation'. The syllabus will further define what is meant by the learning objective.
- Set up a test schedule to mark the trainee's performance of the test items during routine work.
- Agree a standard setting system.

Competency setting is an attractive method of planning curriculum and assessment in practical anaesthesia, but there are problems in deciding what level of behaviour should be observed to assess competence. Most anaesthetic activities are very complex, and breaking them down into chunks to assess separately does not help in deciding if the overall process is competent.

360-degree observation

A more extended form of in-training assessment is one in which everyone who will have an opinion is asked about the candidate. This is called a 360-degree observation. It is now used in the UK as part of the General Medical Council's investigation of doctors against whom there have been complaints of clinical incompetence. It is important that consideration is given to the value of comments received from each team member. Only comments relating to areas of behaviour that the member is qualified to make should be heeded. Criticisms of clinical practice coming from team members who do not understand the judgements involved are not valid. The questions asked of team members should be targeted on their role. Hearsay cannot be accepted. Specific examples of good and bad behaviour must be given. This form of assessment is very powerful in discovering problems in real work. Its uncontrolled and subjective nature requires that there is follow up with 'triangulation' using other methods of assessment aimed at confirming the observations.

Portfolio-based assessment

Keeping a portfolio is revealing of attitudes because the materials in the portfolio must be selected, and much can be learned from what is left out. The text should also include reflections on the materials, and these can form the basis for discussions with tutors. If a trainee is found to have a problem in the affective domain (over or under-assertiveness for example) they can be asked to include incidents that relate to this in their portfolio. They can for example document confrontations and reflect upon how things might have been better. Portfolios are discussed in detail in Chapter 9.

Near-miss analysis

It seems reasonable to believe that poor practice leads to near-misses, and that these can be measured by review. In anaesthetic practice near-misses are called critical incidents. Anaesthetists universally understand the term in this context. Confusion may arise because in the education literature 'critical incidents' are particularly good or bad events during teaching and learning, and 'critical incident analysis' is a methodology for evaluating the progress of learning. Leaving aside the formidable difficulty of coupling career progress to a system of self-reporting of mistakes, there remains a serious flaw in the use of critical incident analysis for competence measurement. The rate of incidents is too low to be able to provide data of significance. A cluster of serious incidents might give cause for concern, but unquestionably this effect could be due to bad luck. Having no critical incidents to report does not mean that a practitioner has shown adequate competence. A recent report using a simulator showed that trainees consistently under-report critical incidents.[8]

Commonly, only the anaesthetist concerned knows that a near-miss has occurred. Boyle showed that a personal profile of cases and critical incidents was useful in directing continuing education.[9] There is a risk that only the honest will be investigated. Critical incidents *should* be reported. Clusters may indicate bad practice; corporate or individual; and there is a need follow up scrutiny of an anaesthetist's practice after a near-miss. Analysis of near-misses has been shown to be associated with a greater rate of reporting of events, and it was possible to identify the sources of some unsafe practices.[10] Critical incident review is usually part of the audit procedures of an anaesthetic department.

Scoring performance

Assessments of clinical performance are often made using rating scales. A rating scale allows the marker to allocate the trainees performance at a point along a continuum from good to bad. Rating scales may have no descriptions relating to the points on the scale, or each point along the scale may be associated with a clear description of what can be expected of a trainee performing at this level. Such descriptions are called anchors. In the context of scoring clinical performance scales with anchors should be used. For formative assessment a rating scale should not have many points on it. Small distinctions between levels of performance do no have much meaning, and are difficult to observe and justify. A rating scale can have just two points, good enough and not good enough. Such a scale allows limen referenced marking of performance. Experts are able to make the simple distinction between satisfactory and unsatisfactory performance just by watching. If the marks are to be used for a 'high-stakes' formative assessment it may be challenged by an appeal, and even by legal action. The marking guidance will then be subject to close scrutiny, and limen referencing will be much easier to explain and defend. For formative assessment, trainees will need more information than a bald pass/fail assessment and a rating scale with more points may be preferred. Such a scale should still be provided with succinct anchors.

The rating scale should include scores for a number of aspects of work, and should include scores for knowledge, psychomotor skills and affective elements.

Two types of learning in medicine:

Factual knowledge:
- Taught in lecture room
- Learned in the library
- 'Knows'
- Tested by MCQs
- Organised for recall in exams

Knowledge of how to do:
- Taught during work
- Learned in workplace
- 'Knows how'
- No formal tests
- Organised for recall during work

A checklist can be used to mark competencies. The competency based training scheme will have broken the work down into a multiplicity of components. Many of these can be observed and marked off on a list. Checklists of this type are very useful for recording the fact that the trainee has completed each necessary stage of a complex piece of work.

Triangulation

Triangulation describes the process of seeking for evidence of competence by using groups of tests that probe the same ability. In the case of clinical difficulties due to poor communication skills evidence could be sought by practice observation, 360-degree observation and ITA. Problems with skill may be detected in these areas but also by critical incident and accident review and audit of outcome. Triangulation is essential in drawing conclusions about major issues, such as dismissing a doctor or planning remedial targeted training.

Assessment in everyday practice

Consultants must learn what to look for when they are making an in-theatre assessment, whether for a daily ITA, periodic reporting or contributing to a 360-degree observation. The best way to develop a policy is by local faculty development meetings. Trainees must then be told what consultants will be looking for when they make assessments.

When a consultant estimates a trainee's clinical performance he or she is exercising their professional judgement about the trainee. We must ask ourselves whether such an approach can be justified for summative assessment. Is it fair, valid and reliable? Consultant opinion is already the way in which day to day training decisions are made. The consultant needs to allow the trainee progressive independence, and it is upon this decision that the safety of the patient depends. When the consultant fills in an assessment form relating to a trainee he or she is recording a judgement that reflects the way training has been progressing. Questions about the performance that the consultant will need to answer are listed in the box.

In the case of periodic reporting, consultants and trainees need to work together often enough for the consultant to form a proper impression of the trainee. The allocation of trainees to work does not usually take this into account. It would be a good idea for departments to make regular placements of trainees and consultants, and this would often mean adopting new work practices in the department.

How to make an assessment by practice observation

In a practice observation a consultant works with a trainee for the purpose of assessment, and scores and records his or her performance during the course of work. The

The consultant should:

1. Let the trainee plan and conduct anaesthesia.
2. Let the trainee make decisions about list organisation, etc.
3. Ask the trainee to explain their actions from time to time.
4. Watch the conduct of anaesthesia carefully at key times.
5. Note examples of particularly good or bad decision making, skill or behaviour.

How to make an assessment by practice observation:

- What was the trainee's general approach to work? Were they well mannered? Were they punctual? Did they organise the work in a businesslike manner?
- How did they perform in the theatre team? Did they communicate properly? Did they take the lead when necessary and co-operate well under direction?
- Did they show appropriate qualities in the conduct of work? Were they vigilant, cautious, and observant? Were they properly decisive?
- Was the trainee's practical work skilful? Did their practical procedures succeed? Was their complication rate acceptable?
- Did the trainee seem to use appropriate knowledge in the course of work? Could they explain the knowledge base that was guiding their choices?

observations take place on an operating list that has been chosen as being appropriate to what is being assessed. Observations can be of work in general, or they can be targeted on a particular activity, such as communication skills or vigilance.

This approach produces a score that tries to reflect how the trainee has performed. It is more structured than assessment by use of consultant reports. The consultant is specifically observing a repertoire of theatre behaviours, and scoring them directly.

The consultant needs to learn the skill of assessing trainees by observing their actions as they work in theatre. The principal focus of observation should be the trainee's skill, judgement, ability to problem solve and professional behaviour.

Preliminary discussion

The consultant should discuss the work with the trainee. The trainee should be asked to summarise what he intends to do. The consultant should take particular note of the trainee's plans for each case, in order to observe and ask about any changes to these plans. The consultant should give consideration to whether the trainee has planned the work properly, and whether he has foreseen the possible difficulties. The consultant should ask the trainee if there are aspects of the work that he is not happy with.

Work observation

The consultant should observe the trainee work. The trainee should be asked to jot down things that they are pleased with, because they went well, and things that they regret, because they went badly. Particular attention is paid to the trainee's method when approaching the times of high work intensity. These times, when there are predictable episodes of clinical decision and judgement, provide good opportunities for assessment. The trainee should spontaneously explain his thinking at key times during the list. This continuous explanation is important in making the trainee's thinking manifest to the consultant.

The consultant notes examples of good and bad practice during the work. At the end of the session these are then used as the basis for discussion with the trainee.

Questions to the trainee during work should help the consultant to decide if the trainee is making decisions rationally. They should not carry a subtext of criticism. Trainees should be asked to explain their reasons whenever they have deviated from their plan.

Review

The trainee should be asked how they think they feel the day has gone. They should be asked to discuss the things they think went well, and the things they think went badly. The consultant should draw them out, not tell them. Each example of particularly good or bad practice noted by the consultant should be explored, and the trainee should be asked about the thinking that was underlying his decisions at these times. The trainee's list of critical events should be discussed.

The trainee should be asked to assess him or herself. The consultant should guide the trainee, but should not at this stage disagree categorically with the trainee's grading of his or her performance. The consultant should then draw the trainee's attention to any assessments with which he does not agree. These should be discussed in depth and, if possible, a consensus reached. The trainee may persuade the consultant that the consultant is wrong.

Finally the consultant should evaluate the trainee. This logical approach to watching the trainee at work can be used either as a formal assessment, or on a daily basis for consultants to reflect on their impression of their trainees.

Assessing affective performance and professionalism

Unacceptable behaviour in the clinical workplace is a frequent cause of difficulty in training and is a cause of suspension and dismissal of career grade doctors. Traditional assessment is not much use for evaluating affective performance. There are however a number of less conventional instruments that can be used. As with all assessment it is necessary to define suitable performance for the guidance of the trainee. Unfortunately there are many consultants who demonstrate unsuitable performance and unprofessional behaviour, and trainees will often protest that their conduct is no worse than that of their role models!

The assessment of behaviour is particularly fraught as the criteria are unclear and the trainee will often say that the consultant is biased against them. All ITA must be open (it must not be anonymous) and behavioural assessment in particular. In my own training scheme it was feared that consultants would stop making valuable behaviour assessments when the trainee knew who had said what. In fact there has been no fall off in reporting bad behaviour, though nowadays, good behaviour is more likely to receive praise.

The in-training assessment form can include specific questions relating to affective components of practice. Punctuality, relationships with patients and staff, ability to organise the progress of work and the ability to communicate clearly with patients are the sort of behaviours that are usually very obvious to a bystander. The assessment form should ask for such information specifically, as it may not be volunteered. Often an assessment form allows the consultant to venture comments and it is amazing how often the same issue will be raised by a number of trainers. Not only have they not colluded in this, but often they have been reticent about voicing their reservations and are relieved to find that they have not been alone in their feelings.

Developing scoring systems for the assessment of clinical performance

In order to assess clinical performance it is necessary to know what its constituent parts are. Most rating scales that are currently in use have not included a systematic investigation of the behaviours that should be observed and scored. What do anaesthetists actually do in the operating theatre? Greaves and Grant[11] used a modified 'Delphi' technique to identify the qualities of practice.

A group of anaesthetists representing many aspects of anaesthesia, and with a variety of experience, were asked to describe the qualities of competent practice. These descriptions could then be reduced to a total of sixteen behaviours that the team felt were always part of competent practice. Such 'qualities' of practice are described as meta-competencies. For each quality the experts then defined behavioural anchors — descriptions of satisfactory and unsatisfactory behaviour. This assessment scheme focuses on general behaviour rather than the individual technical aspects of the anaesthetic task. The advantage of this approach is that the expertise of the supervisor can be relied upon to know precisely what the learner ought to be doing, thus avoiding long lists of specific competencies. The consultant watches the trainee work in a specified situation. If the performance is, in the judgement of the observer, adequate, the consultant need only use the mark sheet to record that the overall performance was

Behavioural anchors for 16 markers of competent practice[11]

	Lacks the necessary quality of behaviour	Appropriate behaviour	Over-expresses the quality of behaviour
1 Knowledgeable: practice is based upon appropriate understanding.	Ignorant: Decisions are not based on sound understanding. Holds mistaken opinion of facts.	Knowledgeable: Practice is based on a sound understanding of principles and facts.	Over-knowledgeable was not recognised.
2 Skilful practice: shows assurance in performing practical tasks.	Unskilful: Clumsy. Ham fisted. Rough. Will undertake procedures without having attained expertise.	Skilful: Shows fluency and expertise in performing practical procedures.	Over-skilful was not recognised.
3 Perception: noticing events as they occur.	Unobservant: Shows delay in noting important events and understanding their significance.	Perceptive: Notices important events immediately and realises their significance.	Lacks discrimination in perception: Notices minor events and over-emphasises their significance.
4 Confidence is the quality of assurance with which work is conducted.	Underconfident: Reluctant to perform tasks within his competence because of lack of understanding of his own abilities.	Confident: Assesses situations and having identified the necessary abilities to carry out a task and compared them his own, will carry on if appropriate.	Overconfident: Attempts tasks beyond his capability by failing to match his own abilities to those required by the clinical solution.
5 Cautious: taking care.	Over-cautious: Fails to take appropriate action because of inaccurate perception of relative risk.	Prudent: Evaluates situations with assurance, identifies hazards, draws up escape routes and institutes prompt action when indicated. Knows his own limits.	Reckless: Proceeds directly to action without thoroughly evaluating the situation.
6 Vigilant: keeping alert for problems.	Over-watchful: Over attentive to inconsequential detail. Pernickety. Too fussy. Allows the search for problems to hinder the progress of anaesthesia and surgery.	Vigilant: Regularly reassesses the clinical situation. Always alert to the possibility of difficulty. Notices problems swiftly.	Inattentive: Does not observe the clinical problems often enough. Does not notice when problems may be arising. Slow to to respond to signs of trouble. Unobservant. Easily distracted.
7 Anticipatory: the quality of thinking and planning ahead.	No anticipation: Does not plan for impending problems. Does not see the possibility of trouble until it arrives. Does not look ahead to anticipate events. Often surprised by events.	Anticipation: Realistically evaluates impending difficulties. Looks ahead and anticipates the future consequences of present events.	Sees too much: Constantly and unrealistically expects serious problems. Always on the edge of his seat.
8 Flexible: prepared to change a plan if circumstances demand it.	Rigid: Does not change plans when presented with evidence that demands this.	Flexible: Constantly re-evaluates decisions as new evidence is presented.	Changeable: Keeps changing his mind without good reason.

(*continued*)

Behavioural anchors for 16 markers of competent practice[11] (*continued*)

	Lacks the necessary quality of behaviour	Appropriate behaviour	Over-expresses the quality of behaviour
9 Responsive: sensitive to what is going on around them.	Unresponsive: Does not appropriately respond to situations or communications. Seems isolated from events. Not part of the team.	Responsive: Takes note of what is going on in the operating theatre. Integrates properly into the theatre team. Attends to what is happening and acts when needed.	Always fiddling: Responds to all communication relevant or not. Easily distracted. Wants to do everyone's job as well as his own.
10 Fluent practice: has progression and continuity.	Hesitant: Having decided on a course of action the trainee is hesitant in carrying it out. A vague vacillating performance. Dithering.	Fluent: The work makes consistent smooth, careful progress. Skilful. Dextrous.	Showy: Skilful and slick at the expense of care and caution. A showman.
11 Decisive: taking timely, purposeful action.	Indecisive: Cannot make up his mind. Slow to respond to cues because he cannot decide on a line of action.	Decisive: Makes good decisions after proper consideration of alternatives.	Quickly proceeds to action without full consideration. Jumps to the first course of action that occurs to him.
12 Communicative: articulating ideas and intentions to others.	Poor communication: Fails to communicate even when the clinical situation demands it. Conveys muddled information.	Communicates well: Identifies priorities in communication. Conveys information in a relevant, clear and concise way.	Overvoluble: Communicates everything. Distracts colleagues from their tasks by inappropriately offering information.
13 Organisation: maintaining system and order.	Disorganised: Rushes about without making progress. No system. Jumbled. Muddled. Does not plan.	Well organised: Shows a systematic approach. Orderly. Planned.	Over-coordinated: Fanatical about organisation. Rigid organisation in minute detail. Bogged down in planning.
14 Manner: the conduct of relationships during work.	Unmannerly: Disregards the feelings of others. Insensitive. Rude.	Manner: Behaves in a manner that is acceptable to colleagues and patients.	Diffident: Sacrifices effectiveness by being over polite.
15 Assertiveness: describes the ability to take the lead.	Underassertive: Does not make his presence felt. Not in command at a crisis. Does not give a lead.	Appropriately assertive: Firm and disciplined in command. Understands where to take a lead and is comfortable in that role.	Overassertive: Bossy. Disruptive. Aggressive. More concerned with keeping other people out of what he sees as his business than with co-operating to get a good job done.
16 Management: describes the capacity to organise work and lead a team.	Disorganised: Does not organise others as a team. Fails to organise the work.	Management ability: Plays an appropriate part in planning and managing work.	Overmanages: Concentrates on organisation at the expense of practice. Does everyone else's job.

satisfactory. If, however, the consultant is unhappy with the trainees work, the check sheet of performance with behavioural anchors is used to provide a description of what was wrong. This description can be used for summative assessment, but is also valuable formative assessment to feed back to the trainee.

Forrest et al. used a classical Delphi technique to compile an inventory of key technical elements during anaesthesia. The inventory was derived from textbooks, journal articles and the experience of the researchers, and the panel was then asked to score each element. The scoring system that was so developed was then tested by observation of standard scenarios in a simulator. Reasonable inter-rater reliability was seen, and a group of novice anaesthetists was shown to have improving scores with experience.[12]

Glavin and Maran have advocated the use of cognitive analysis in identifying the important elements of the anaesthetists work.[13] Fletcher et al. used this approach to define the non-technical skills of anaesthetists.[14] In this technique interviews and observations are used to unravel the behaviours used in the course of work. These elements of behaviour can then be used to develop an assessment tool. The taxonomy developed by Fletcher is used throughout this book to describe the non-technical skills of anaesthetists.

Monitoring progress with practical procedures

The logbook records the numbers of procedures and whether they were successful. It is quite difficult to tell whether the learner is making satisfactory progress from raw logbook data. CUSUM (cumulative summation of scores) has been used to analyse progress in learning. An acceptable failure rate for the procedure is determined and each successive procedure is plotted on a graph. The profile of the plot shows when the learner has reached the acceptable failure rate.[15] Harrison has described an alternative sequential technique.[16] A grid is constructed by analysis of the performance of a group of learners. Each cell is marked with the percentage of trainees that will pass through it. The individual learner's attempts are then plotted through the grid — up for a failure and along for a success. Using this scheme it is possible to see whether learning is proceeding well, or whether performance is well below expectation. This technique compares the individual's performance with that of the group. It is norm referenced, whereas the CUSUM technique described above is referenced against a criterion determined by the expert group. Unfortunately formal progress analysis schemes are not available off the peg and departments would have to calculate their own grids.

References

1. Ferguson C, Barnes A. Residency program criteria for the completion of clinical competence reports. *Anesthesia & Analgetics*. 1991;72:S86.

2. Viets JL, Foster SD. A clinical evaluation system for anesthesiology residents. *Journal of Medical Education*. 1988;63:463–466.
3. Gray JD. Primer in Resident Evaluation. *Annals Royal College of Physicians and Surgeons of Canada*. 1996;29:91–93.
4. Sugiura Y, Miyamoto E, Harada J, Goto Y, Takahashi K. Evaluation of the training of the anesthesiologist. [In Japanese] *Masui: Japanese Journal of Anaesthesiology*. 1996;45: 766–768.
5. Parenti CM, Harris I. Faculty evaluation of student performance: a step toward improving the process. *Medical Teacher.* 1992;14:185–188.
6. Rhoton MF, Barnes A, Flashburg M, Ronai A, Springman S. Influence of anesthesiology residents' noncognitive skills on the occurrence of critical incidents and the residents' overall clinical performances. *Academic Medicine.* 1991;66:359–361.
7. Jago RH. Confidential professional reports. A method of assessing the career progress and prospects of anaesthetic Senior House Officers. *Anaesthesia.* 1989;44:1321–1328.
8. Mackenzie CF, Jefferies NJ, Hunter WA, Bernhard WN, Xiao Y. Comparison of self-reporting of deficiencies in airway management with video analyses of actual performance. *LOTAS Group. Level One Trauma Anesthesia Simulation*. 1996;38: 623–635.
9. Boyle RK. Quality assessment in personal anaesthetic practice. *Anaesthesia and Intensive Care*. 1993;21:331–334.
10. Over DC, Pace NA, Shearer VE, White PF, Giesecke AH. Clinical audit of anaesthesia practice and adverse peri-operative events at Parkland Memorial Hospital, Dallas, Texas. *European Journal of Anaesthesiology*. 1994;11:231–235.
11. Greaves JD, Grant J. Watching anaesthetists' work: using the professional judgement of consultants to assess the developing clinical competence of trainees. *British Journal of Anaesthesia*. 2000;84:525–533.
12. Forrest FC, Taylor MA, Postlethwaite K, Aspinall R. Use of a high-fidelity simulator to develop testing of the technical performance of novice anaesthetists. *British Journal of Anaesthesia*. 2002;88:338–344.
13. Glavin RJ, Maran NJ. Development and use of scoring systems for assessment of clinical competence. *British Journal of Anaesthesia*. 2002;88:329–330.
14. Fletcher G, Flin R, McGeorge P, Glavin R, Maran N, Patey R. *Final report: Development of a behavioural marker system for anaesthetists' non technical skills (ANTS)*. [Grant report for SCPMDE, project reference RDNES/991/C]. Aberdeen, UK: University of Aberdeen, 2001.
15. Kestin IG. A statistical approach to measuring competence of anaesthetic trainees at practical procedures. *British Journal of Anaesthesia*. 1995;75:805–809.
16. Harrison MJ. Tracking the early acquisition of skills by trainees. *Anaesthesia*. 2001;56: 995–998.

17

Giving feedback and monitoring progress

Berend Mets

Anaesthetists are often confused about the meaning of the various terms used in the processes of evaluation and discussion of the learner's performance. Assessment, evaluation and appraisal all depend on a supervisor developing an opinion of the learner's clinical performance. Feedback is the day-to-day sharing of these responses and it underlies all the other processes. It must be done and it must be done well.

Feedback is the Breakfast of Champions!

Learners may prefer feeding to feedback.

Feedback

Feedback is formative. It is to be used by the learner to direct their learning. Its first function should, whenever possible, be praise and encouragement. Trainees can spend many weeks working with a variety of supervisors, performing well and receiving no praise. They do not know how well they are doing. Even when it is really obvious to the consultant that the trainee is doing well they may be going through agonies of insecurity and self-doubt. When the trainee is having problems the consultant should dwell first on the positive and when raising

Avoid undirected praise or criticism.

Howd' I do?
You did great!

Take every opportunity to praise and encourage the trainee.

concerns they should always try to help the trainee by suggesting what can be done to improve matters. Consultants must remember the devastating effect that their criticism will have, even on a trainee who appears to be very resilient.

Teaching and learning should be interesting and they should be fun. Always try and assess whether the trainee enjoyed the day. Some trainees have difficulty with the theatre learning situation because they cannot combine formal learning with looking after patients. This is a behaviour that should be respected; though occasionally it is the mark of a trainee who is over anxious in clinical situations. Sometimes the trainee will be dissatisfied because the consultant tried to teach material that was inappropriate to their stage.

Feedback should be goal directed, descriptive, and immediate. It should provide information about the performance of a trainee, presented in a non-judgemental fashion without reference to the inherent value of the individual. Feedback is key to learning in the work place as it corrects misconceptions and clarifies puzzlements. In the absence of feedback trainees may learn bad habits and, more worryingly, may misinterpret or add inappropriate weight to a raised eyebrow or an (apparently) disapproving glance. In addition instead of feedback from a consultant they will rely on their own interpretation of their performance with ultimate, subsequent, loss of receptivity to any feedback and even arrogance with respect to their own abilities.[1]

Feedback should be:

- Immediate or at least as timely as possible.
- Start with the positive things.
- Preferably in a neutral environment (do not do so in the heat of the moment when you are upset).
- Based on first-hand observation.
- Specific and descriptive.
- Non-judgmental.
- Directed towards behaviour not personality.
- Checked with the recipient.
- Problem solving.
- Based on suggestions rather than prescriptions.

Feedback is often avoided, for fear of 'criticising' the individual, or due to lack of time or a suitable place to do it. In addition many tutors do not sufficiently observe, assimilate and process the necessary information to make a sufficient assessment of the clinical performance of an individual to support appropriate feedback.

Have the trainee paraphrase the feedback and invite questions and discussion.

Further, trainees may not wish to have feedback preferring 'feeding' a confirmation of their own self-image. This can occur particularly where the so-called 'vanishing feedback' is given. In America this is the non-specific end of the day comment, "You did great," usually in response to the trainees question, "How did I do?" This inevitably results in the response "Fine" and does not address real issues in order to avoid 'upsetting' the trainee.

Trainees should be encouraged to seek feedback about their performance by asking, "How might I have done this better?" Or "How would you have done this differently?" This can hardly elicit a response such as "Fine."

Practical feedback in theatre/OR

Feedback don'ts:

- Don't use generalisations that refer to trainees' diligence, efficiency, concentration, prioritising and anticipation without using specific examples.
- Don't feed back interpretations or assumed intentions — focus on actions.
- Don't praise i.e. "You're terrific" unless you give a specific reason so that the trainee knows what was terrific, e.g. "Your ability to keep your cool (when the aortic cannula fell out during cardiopulmonary bypass), calm the surgeons, support the blood pressure with vaso-active agents and restart ventilation at the spur of the moment was terrific." (Well wasn't it?)
- 'Embarrassing praise' and 'Humiliating criticism' should be avoided.

OK so you, the reader, are now convinced that 'Feedback is the Breakfast of Champions' but how should you do this in real life.

Let me give you a hypothetical case and suggest a few pointers on how you might proceed with a feedback scenario with two different types of trainees, the responsive and the not so responsive types.

Try not to loose your temper!

You may feel better but the trainee will focus on your anger rather than the mistake he or she is making.

Case scenario

You will feed back on a simple airway management, intubation, and start ventilator scenario.

The trainee starts mask ventilation after induction and the administration of a muscle relaxant but does not introduce an airway although the stomach is slowly being inflated.

You hand him an airway.

He goes to intubate, using the laryngoscope he lacerates the lip, intubates, watches the chest for inflation but does not auscultate, connects and starts the ventilator, but does not evaluate the tidal ventilation or frequency of ventilation.

Obviously not to comment would be unacceptable. How can one go about feedback, while ascertaining ego-protection and not getting into a confrontational situation?

Firstly cool off, if you are upset about the stomach inflation and the lacerated lip. Then, during a quiet phase of the case and out of earshot of the surgical team, after you have said that you will monitor the patient (so that the trainee can concentrate) you could ask the following questions of the trainee.

- What are you pleased with?
- How could you have done it better?
- What are you not pleased with?
- What would you suggest in the future?

Note that the approach here is to have the trainee diagnose the problem with you thus creating a 'teaching moment' where the individual will be receptive to whatever feedback you might give.

It is hoped that the trainee will pick up the glaring omissions: lack of an airway, non-auscultation of the chest post intubation, non-patient specific adjustment of ventilator settings. Having done so, we hope that he learns to do these in the future and is more careful about intubation to avoid lacerating the lip.

The less responsive resident who does not identify the issues may require you to take the following approach:

- Describe to him what you saw.
- Ask: "What were you thinking?"
- Ask: "What goals would you like to achieve?"
  (Provide them if he or she hasn't got any: Avoid stomach inflation, atraumatic intubation, auscultation after inflation, individualised ventilatory parameters.)
- Ask: "What would you do differently in the future when performing this sequence of airway management?"
- Make suggestions if he does not provide them.
- Ask him to paraphrase the feedback.

Feedback for formal monitoring of progress

We will now consider the more formal monitoring of progress for which feedback is used. There is some confusion about the meaning of the various terms used in these formal settings. The terms that need to be defined and related are *evaluation, assessment, supervision* and *appraisal*.

There is more agreement about the educational processes; of which there are three.

The first process is giving the trainee immediate information about their performance. This is feedback and we have already discussed it in some detail. A learner benefits from guidance to understand the work and their learning. If the learner is working with only one or two teachers for an extended period then they can provide this support

The three processes for monitoring training:

	The learner's role	**The trainer's role**
Process 1. Mentoring or Supervision	Discusses work and progress with their nominated trainer. May also discuss more personal matters if the learner requires. The discussions are confidential.	To support and aid the trainee in achieving their goals.
Process 2. Appraisal — UK Feedback — USA	A formal discussion of progress including any relevant documentation. Any disclosures of process or outcome must be agreed by the trainee.	To review the trainee's experience of training and provide advice on progress. To discuss with the learner any in service assessments. To note any problems the trainee has had with the scheme or individual trainers.
Process 3. Assessment — UK Evaluation — USA	To provide evidence of adequate progress. The outcome will be known to all relevant trainers.	To compare the trainees evidence of progress with the required standards for further advancement.

and the discussions can take place in the course of work. In anaesthetic training schemes and residencies the working relationships are often fragmented and many departments will appoint a mentor or personal tutor to help each trainee.

The second process is more structured. From time to time the trainee needs to formally review how they are progressing. This process can take place by discussing all aspects of their progress with a well-informed consultant anaesthetist. The discussion will require the trainee to provide information about what they have done and this will probably include written evidence. It will include information about the clinical work done and will include the portfolio of learning if there is one. The discussion will include the performance of the training scheme as well as the progress of the learner. This process is best carried out by a trainer who is neutral, not the trainee's mentor nor an office bearer of the training scheme who will be responsible for the next process. Confusingly, in the USA this second process is often called feedback, though the term is also used in its less specific form to describe the two way process of exchanging information about learning as above.

The final stage is that of measuring progress. The standard is either criterion referenced (must have done 20 cardio-pulmonary bypass cases), or limen referenced (his trainers have scored his clinical performance as adequate). The crucial difference about this final process is that the learner can fail. In the first two processes the learner can set their own targets and evaluate their own achievements and failures. In this final process the standards are entirely defined by the trainers.

How are these three processes used and described in anaesthetic training?

Appraisal is the term used in the United Kingdom for the second stage of monitoring progress.[2] In the USA this is usually called *Feedback* and is conducted by a Faculty Advisor assigned to the resident in training. In both systems it is a formative process, in which information (not judgement) is provided to the trainee, career goals and objectives established in a regular, confidential discussion and a training agreement established. The latter is an agreed outcome of the meeting and is recorded. Feedback/appraisal should not be confused with assessment/evaluation.

Assessment/Evaluation and Appraisal/Feedback compared:

	Assessment/Evaluation	**Appraisal/Feedback**
Purpose	To measure performance against external, agreed criteria.	Describe current performance. The intention is to arrive at an agreement with the trainee.
Information available	Test scores, standardised performance reports, ratings by supervisors.	Portfolio and evidence edited and volunteered by the learner.
Methods used	Knowledge tests, professional exam performance, standardised clinical observations, oral exams, objectively structured clinical examination.	Extended interviews and peer group reports of performance.
Outcome	Summative assessment: May control movement through a training scheme and ultimate qualification.	Formative: Enables modification of experience, training and learning plans to accommodate the individual learner. Cannot be failed. The discussion is private but an agreed statement of outcome may be public.

For the third process of monitoring training in the USA the terms *evaluation* and *assessment* are used interchangeably, whereas *assessment* is the term more usually used in the UK. These terms are used to describe the process of judging the progress of trainees based on clinical competence, knowledge base and professionalism.[3] By nature this is summative, comes after the fact, and is a judgement, usually by the faculty/consultant involved. Assessment has also been defined as "measuring progress against defined criteria based on the relevant curriculum".[4]

Appraisal

To reiterate, appraisal is a formative process where a neutral consultant meets with a trainee, on a regular (though not frequent), confidential basis. It is used in conjunction with the trainee, and with knowledge of previous assessments, background, aptitudes and career plans, to determine a course of action to realise the necessary education and training goals. Done effectively the appraisal may be a positive motivational experience for all involved.

The appraisal should be performed three times during a post, at the start, midway and the end.[5] The outcome of the appraisal should be a jointly agreed learning plan, which should be written and become part of the appraisal record.

Remember, however, that the discussion should be confidential and should be non-judgemental. Hence assessment (a judgemental process) should probably not be done at the same time as the appraisal as it is difficult for the consultant to switch from a summative process (judgemental) to a formative process (information giving).

Who should do the assessments and who the appraisals?

Some consultants may do the assessments (in theatre, the ITU, the pain clinic) while one (in a small department) may do the appraisal, although there should be an opportunity for the trainee to have appraisals performed by another consultant if requested. There is something to be said for consultants who have an aptitude for performing appraisals to be assigned to do so. There is no doubt a learning curve. Further, the individual(s) landed with this most delicate of tasks should have the respect of his/her colleagues/peers and certainly the trainees. Some training schemes in the UK allocate more importance to one of the appraisal meetings each year in this case it is done by one or more 'neutral' consultants. In the US, appraisal is termed feedback and is usually conducted by a faculty advisor assigned to the resident. In a review on the subject, this is performed biannually in 40% and quarterly in 39% of programs with more frequent feedback if performance problems are identified.[6]

Why do appraisal?

The cynic might say because it is required, but there are important reasons for appraising trainees and giving feedback on their performance, as trainees should know what they are supposed to do and how well they are doing it. This will help the trainee identify their own, strengths, weaknesses, and educational needs.

How should appraisals be performed?

The consultant

The consultant should allocate sufficient time, and an appropriate, private location needs to be used.

The consultant should be prepared for the meeting, having reviewed the trainee's assessments, CV and gleaned whatever background information from the faculty tutor, as well as other consultants that might be of use. This especially true if the appraisal is of a trainee who has unsatisfactory progress (but more about that later). The consultant would do best to have a plan of action and a desired outcome of the meeting, but must be able to be flexible in the course of the meeting if new insights are gained.

The basic skills needed of the consultant to carry out an effective appraisal relate to the ability to feedback, listen, support, counsel and ask appropriate questions.[7]

At the start of the appraisal the Consultant should put the trainee at ease, but be aware not to send mixed messages. For example, if the resident's performance is unsatisfactory, the consultant should not give the impression that this is not a serious issue, which needs to be addressed. Nevertheless, the meeting should be seen in a positive light and the consultant should attempt not to be judgmental.

He/she should state the facts as the consultant sees this, referring to specific assessments and specific instances (time and place) as addressed there.

- Then the trainees should be asked to comment on this and given the opportunity to add what they have done well and how they did this.[4]
- The consultant should reinforce this where he/she feels this is a correct analysis.
- The consultant might ask what could be done differently and how.
- The consultant then re-enforces the correct analysis.
- The consultant will review the experience log, assess the level of knowledge of the trainees, discuss career goals and in the light of all of this work together with the trainee decide on an action plan for the future.

A good suggestion would be to ask the trainee to paraphrase the most important points of the discussion.

The action plan should be written down and certainly in my own Residency program in the USA this is filled out on a purpose built form (Faculty Advisor Feedback Form) with a space for the resident to countersign.

The trainee

Trainees should prepare for the appraisal, by ensuring that logbooks and CVs are up to date as well as by reflecting on his/her career goals.

- Be on time, time is precious.
- Remember this is not a job interview.
- Understand that while this appraisal is difficult for you, it may not be so easy for the consultant either.
- Understand that the objective of this process is to help you identify weaknesses and strengths in yourself but also to give honest feedback about available resources (books, computers) and the teaching that is performed in theatre, the ITUs and pain clinics.
- Do this without assigning blame or without rancour, i.e. do not blame your lack of progress (if this is the case) on lack of resources. Remember you are a post-graduate trainee and your education is ultimately your own responsibility.
- Be aware that the consultant will (at least roughly) use the guidelines above and attempt to help the process along.
- Be prepared to establish with your appraiser realistic objectives for your training and education. Do not promise what you know that you cannot deliver; it will only lead to disillusionment at the next appraisal session.

Underperformance/remediation

A most difficult problem for the appraiser is dealing with the 'unsatisfactory' or underperforming resident.

The first problem that the appraiser faces is making a diagnosis of the problem and categorising the problem. They must then plan and monitor a course of action to be followed and ultimately decide whether 'termination' from the program must be advised.

The process we follow in our residency program may be illustrative and is summarised briefly below, after which I will make a few general observations.

In our Residency Program in the USA we categorise all resident performance in terms of three competencies (*Academic, Clinical* and *Professional*) and set goals and objectives for these as well as defaults which would allow us to put a resident on what is known as 'remediation'.

So for *Academic competence*, residents need to achieve scores on a number of validated national anaesthesia examinations so as to be ranked above the 25th percentile. (Please note these are not exit exams). If they do not do so, first a diagnosis is made as to the reason for inappropriate performance (Study habits, motivation, disease, personal problems etc always bearing in the back of one's mind that there is a high incidence of drug diversion in anaesthesia). Then a course of study is outlined, and a faculty teacher monitors this and sets in-house, written examinations to monitor and encourage progress. The resident is taken off academic remediation when scores are reached above the 25th percentile.

Clinical competence is evaluated (assessed) in the operating room, ITUs and pain service for which composite (from a number of faculty grouped by subspecialty) evaluation forms are filled out, reviewed, and filed. These forms have different evaluation categories for which the evaluators mark off whether residents are: *outstanding, above average, satisfactory, marginal* or *unsatisfactory*. Using a predetermined performance policy, which is advertised and is part of the resident manual, if residents receive repetitive *marginal* or *unsatisfactory* evaluations they are put on *Clinical remediation*. In an appraisal interview the evaluations are discussed, and a plan of action outlined and written down and co-signed by the resident and consultant (usually the chair of our clinical competency committee). We have also adapted an article on Clinical Competency Criteria to fit in with our residency program structure and use these written competency guidelines to assess whether the trainee in question is at the appropriate competency level.[8]

Professional competence is judged according to the criteria set for this by the American Board of Anesthesiology.

Overall, we find that if one of the Competencies described above for a specific trainee is impaired we look very carefully at the other two competencies. If two are impaired (and do not improve on remediation) this may be a potential trainee that will need to be referred.

In the UK the system for dealing with under-performing trainees is organised by the relevant university's Postgraduate Dean. At an annual review of training all evidence relating to an individual's progress is considered and a certificate for further training is issued. Evidence will include success or failure in professional examinations, in training assessments by supervisors, the logbook of training and any portfolio of learning.

- A trainee may pass satisfactorily to the next stage of training.
- A trainee may be recommended for extra training in a particular area without having to extend their time in training.
- They may be recommended for special training under a special educational programme. Their training will be extended by this period. At the end of this period they will be reassessed. If they are found still to be under-performing they may be dismissed from training.

Though no figures exist it is the general opinion amongst programme directors that problems relating to professionalism are the most frequent cause of termination.

What is the consultant to do when faced with the unsatisfactory trainee?

Clearly, based on the experience (described above), first a diagnosis of the problem needs to be made. Is the problem, clinical, academic or professional or a combination of these? Then, preferably, based on pre-established criteria, a plan of action should be determined in conjunction with the trainee. This needs to be paraphrased by the trainee and written down and a time and place for follow-up determined.

Clear (or as clear as possible) endpoints should be established for performance. It is also important that if the Consultant feels that there is a high likelihood that the trainee appears unlikely to be suitable to continue to train in anaesthesia that this is communicated at an early stage in the process. The consultant needs support from others in the department in this instance to minimize the potential for recrimination and to ensure that the consultant makes these often very difficult decisions and does not leave it to someone else.

This Chapter has dealt with Appraisal, Assessment/evaluation and Feedback. It should be clear that these are all interrelated. The appraisal process requires of the Consultant considerable skill and motivation to ensure that the trainee can gain the maximum benefit. The trainee should be informed about the process and prepare prior to the appraisal meeting to gain the maximum benefit from this interaction.

References

1. Ende J. Feedback in clinical medical education. *Journal of the American Medical Association*. 1983;250:777–781.
2. Riley W. Appraising appraisal. *British Medical Journal*. 1998;Classified Nov:2–3.
3. Strachan A, Hood G. *Assessment of trainees – A survey of college tutors*. Bulletin 1. London: The Royal College of Anaesthetists, 2000;17–18.
4. Guide to specialist registrar training.
5. Sinclair J. How to maximize the benefits of junior appraisals. *Hospital Doctor*. 2000; 42–43.
6. Rosenblatt MA, Schartel SA. Evaluation, feedback, and remediation in anesthesiology residency training: a survey of 124 United States programs. *Journal of Clinical Anesthesiology*. 1999;11:519–527.
7. Campbell J, Wiliams S. *2 Sides of A4: Appraisal and Feedback*. Occasional Paper. University of Wales College of Medicine, 1998.
8. Madsen K, Woehlck H, Cheng E, Kampine JM, Lauer K. Criteria for defining clinical competence of anesthesiology residents. The Clinical Competence Committee to the American Board of Anesthesiology (Guideline). *Anesthesiology*. 1994;80:663–665.

Section 3

Using simulators for teaching

18

An introduction to simulation in anaesthesia

Ronnie Glavin and Nicki Maran

A simulator is a machine that simulates an environment for the purpose of training or research.

In the next five chapters we shall look at the place of simulation in Anaesthetic education. The number of centres with sophisticated simulators has increased considerably over the last decade, from a mere handful to over a hundred, and simulation stands on the threshold of becoming a routine teaching tool in medicine.

We shall attempt to look at what simulators have to offer as a tool in teaching, assessment and educational research. How can we get the best use out of them? Where can they fit into our training programmes? Should every teaching centre have one? In this chapter we shall look at these topics in general.

Like most things in medicine, simulation is not new and the history of present day simulators can be found in an excellent chapter by Professor David Gaba.[1]

Many different devices called 'simulators' are available and rather than begin to list them we shall attempt to classify them. The most appropriate method in the context of anaesthetic education is to consider which aspect of the environment the device is attempting to replicate.

At the most basic level we have devices that only attempt to replicate part of the environment. These are the part task trainers we use when we want to teach or assess airway skills, venous access, regional anaesthetic catheter placement etc. Computer models that deal with the pharmacokinetic properties of inhalational or intravenous anaesthetic agents fall within this category.

A new development in this area is haptic systems used with or without virtual reality. Haptic (Greek captein to touch) devices aim to give a realistic feel to simulated practical skill trainers. A set of motors in three dimensions give the operator the impression of the degree of resistance when inserting a needle, such as a spinal needle or a suture. The virtual reality part is an attempt to project a visual image of the scene, so that while watching an image of a patient's back (using data from the visible human project), and handling a specially mounted needle, the operator is given the impression of performing insertion of a spinal needle. Haptic systems are of particular interest in surgical training, and anaesthetic developments include local anaesthetic blocks.

Types of simulation device:

- Part task trainers including haptic systems
- Desk top systems
- Low fidelity systems
- Intermediate fidelity systems
- High fidelity systems
- Virtual Reality
- Simulated patients

The next level of complication is to model aspects of human physiology and allow interaction with these. Desktop systems such as Howard Schwid's 'Anesthesia Consultant'[2] or Ty Smith's 'Body'[3] allow the participant to interact with the computer model through a keyboard. Pharmacological models of commonly used anaesthetic drugs are included, so the participant can choose the drug and dose and can see the response. However, all the interaction is done via the computer. The participant obtains information from a screen and interacts by using the keyboard or mouse.

Some advantages of simulation training as compared to real operating theatre teaching:

- No risk to patient.
- Scenarios involving uncommon but serious problems can be presented.
- Same scenario can be presented to trainee many times.
- Same scenario can be presented to many trainees.
- Errors can be allowed without any risk to the patient.
- Simulation can be stopped and then repeated.

The next development is to combine the computer models with a patient in the form of a manikin. The degree of fidelity will depend upon both the level of sophistication of the manikin and the computer models. The ACCESS system, designed by Aidan Byrne[4] and the Laerdal 'Sim man'™ represent the lower and intermediate fidelity models. At the higher end of fidelity are the METI Human Patient Simulator™,[5] the MedSim Patient[6] and the Sophus simulator.[7] The Leiden

Anaesthetic Simulator falls into this category but is not available commercially.[8] The higher fidelity models allow clinicians to interact with the patient as they would in the real clinical environment. They obtain information from the patient, speakers in the manikin's head create the impression of the manikin talking, and physical signs are generated. In turn the clinician can carry out therapeutic procedures on the manikin. Its oxygen saturation and blood gases will be affected by its inspired oxygen concentration, it will respond to drugs and some part task training procedures can be effected. Virtual reality at the high fidelity level is in its early stages. It is used more at the part task device level but the potential to carry out the same functions as the current high fidelity simulation devices is there.

Finally, we have simulated patients. These are real people who are trained to behave like patients with real diseases. They are used mostly in undergraduate education where access to real patients can be difficult.

Most people have some familiarity with the concept of high fidelity simulation as seen in aviation. The fidelity is such that if a pilot can fly a particular model of plane on the simulator, then he or she can fly the real version. It is important that anaesthetists understand that their professional role cannot be so completely reproduced for training and assessment. There are several reasons why this is so. Firstly, it is easier to predict how an aeroplane will react to pilot intervention because we are dealing with the laws of physics, and a lot of data has been generated in allowing the successful application of these laws. The models that 'drive' the simulator are therefore accurate representations of the reality. Secondly, pilots rely on instruments to obtain more of their information than do anaesthetists. Thirdly, aeroplanes will respond in a very similar manner. One 747-400 should behave in the same way as another 747-400. This allows the aviation community to carry out a lot of teaching on flight simulators and to use these devices for assessment purposes.

Patient Simulators are very different. The models deal mainly with the respiratory and cardiovascular systems and the data used as the basis of the models are derived from normal human volunteers. The complexity of human physiology is nowhere near being replicated. Human beings differ vastly even in response to relatively simple interventions, such as the same weight-related dose of intravenous anaesthetic agent. We cannot even effectively replicate many items of information that we use in the management of patients, such as facial expressions, muscle tone and skin colour. Does that mean that we cannot use simulators for learning? Of course not, but we cannot expect to import the culture of simulation from aviation unchanged. We have to use their strengths in pursuit of our educational goals.

In the next chapters we shall look at specific applications of simulation to teaching, but first we shall look at the theoretical educational benefits that can come from simulation, and why simulation can be a very useful complementary tool. A pilot can learn how to fly an aeroplane by practising on a simulator. Anaesthetists, however, will learn best by working with real patients under supervision. We believe that there are good theoretical reasons why simulation can enhance that learning, but simulation should be seen in the context of all the other learning methods available.

Relevant concepts from educational theory

The two main themes of educational theory that are of relevance are the application of adult learning and the principles of experiential learning. These have been discussed in previous chapters and the reader is referred to the relevant sections but we will here expand upon some aspects of these themes.

Experiential learning and simulation

The experience cycle:

- Undergo the experience.
- Reflect on the experience.
- Compare the experience with existing knowledge.
- Test the new knowledge by starting a new cycle of experience.

The essence of experiential learning is that learning is an active process. We do not passively absorb information, but construct knowledge by linking new information and new experiences with what we already know and understand. Learning can therefore be seen as a process of interpretation, integration and transformation of the world as one encounters different experiences. What has happened to me in the past will influence what I expect to happen to me in the future. In this way we try and make sense of the world by fitting things into patterns. This process has been described in terms of a cycle; the experiential learning cycle (Chapter 3).

The first phase of this cycle is to undergo a specific experience. In the context of anaesthesia this will be the management of a patient. This leads us into the second phase of the cycle, reflection. What has happened? What changes have we seen or

How to use the simulator to create ideal conditions for learning in the four phases of the experience cycle:

Phase 1 — Experience. Create episodes of experience rather than using what work offers.

Phase 2 — Reflection. Halt or replay the scenario and so on. Focus more on details than would be possible in the real clinical environment.

Phase 3 — Linking observations to underlying knowledge. Allow the learner to look up source material etc.

Phase 4 — Testing hypothesis. Allow the learner to conduct the experiments that will confirm or refute their hypothesis, even if these would be dangerous to a real patient. We can then let the learner modify the experiment in the light of the findings.

heard? What have we noticed? This is how we come to interpret events. The third phase is where the observations and interpretations are tested against the knowledge one already possesses. Did I expect that to happen? If there is a conflict between our observations and our existing knowledge then we have to come up with possible explanations. In other words we generate hypotheses. The fourth stage is to try out one or more of these hypotheses. This experiment of course becomes the episode of concrete experience of the first stage and so we are back to the beginning of the cycle to go round it one more time.

Good teaching experiences do not necessarily arise during real work and it is not always possible to reflect fully on an experience. It is not easy to ask questions let alone answer them as a learner, when there are other concerns competing for our attention. Simulation allows replay and modification of events and responses. Simulation therefore allows deliberate use of the experiential cycle model.

Principles of adult learning

The second aspect of educational theory that is relevant is an increased understanding of the principles of adult learning. The schoolroom or classroom method of teaching was, and probably still is, the most widely used approach to teaching in medicine. In this style of teaching there is a teacher and group of learners. The teacher sets the agenda in terms of the lesson for that day, the teaching methods and any assessment methods to be used The learner's role in these aspects is passive. These are the methods used in formal lectures in undergraduate courses, or even small group work in examination preparation courses in postgraduate teaching.

Successful adult education programmes have used the new understandings of the specific needs of adult learners to move away from the classroom method of teaching into one in which there is a partnership between teacher and learner and the learners have an input into what is learned and how it is learned. Learners are given more responsibility and encouraged to use that responsibility. These principles are

The specific needs of adult learners:

- In adults readiness to learn is a function of the need to perform social roles.
- Adults have a problem-centred orientation to learning.
- Adults need to know why they need to learn something before commencing their learning.
- Adults have a psychological need to be treated by others as capable of self-direction.
- Adults have accumulated experiences and these can be a rich resource for learning.
- For adults the more potent motivators are internal.

exemplified in the Problem Based Learning approach. In the previous section we saw the value of reflection and this is formalised in simulator training in the debriefing session that occurs after the simulator experience. Debriefing lends itself to the adult education approach. The active participant in the simulated scenario is encouraged to review his or her performance and to select aspects of concern (if any). The experience of both the active participant and others attending the session can be called upon to help interpret what was happening, and to discuss possible ways of management. From this discussion the learner is encouraged to set learning objectives, by identifying areas where performance was not to the desired standard. Adults are more likely to act on these learning objectives if they have set them for themselves, instead of having someone else set them. All aspects of anaesthetic teaching can be modified to incorporate the principles of adult learning, but the opportunity to review one's performance can be a most powerful tool for generating educational objectives.

The general principles of using simulators

Simulators are educational tools. If we are to get the most out of them then we should use good educational principles. In general a session with simulation is in two phases:

- The performance of the learner using the simulator.
- The review of that performance, or the debriefing.

If a scenario is to be 'educational' then two criteria must be satisfied. There must be intention, we have a specific educational objective, and it must be of value to the learner. In the context of anaesthesia this means that it should help progress the learner towards the end of the training programme, or the career grade anaesthetist towards a higher standard of practice.

We choose our scenarios on the following basis.

- *Based on a real incident.* This allows a degree of accuracy when compiling the story and the set of values one needs. It also gives the scenario some credibility. There is a temptation to overload scenarios, especially when using a high fidelity simulator in the clinical area. There is a temptation to use many of the working features of the manikin and this is not necessary to satisfy good educational principles.
- *Scenarios should allow the degree of difficulty to be changed.* The scenario should be sufficiently flexible that one can make it more challenging or less challenging, depending on the performance of the participant. If the scenario is too stressful, if the participant's self esteem is badly damaged then the debriefing part will be completely ineffective. A participant who is 'shell shocked' is unlikely to be willing or able to review his or her performance.
- *Allow choices for the participant.* This helps to broaden the discussion during debriefing, and allows the active participant and others to explore their own experience to a greater degree.

The combination of carrying out a clinically challenging case and then immediately reviewing the performance in a setting that is made as non-threatening as it can be, gives learners the opportunity of as objective a review as is possible. Using video recordings of the scenario adds to that objectivity, as the subjective recall of the learner and the subjective recall of the teacher can take second place to the neutrality of the recording. Our experience is that it is easy during a scenario to remember the wrong actions taken. It is not normally so easy to remember the elements of good practice. These become very apparent during play back of the video recording, and so our debriefings last at least as long as the scenario, and are more likely to last twice as long. It should go without saying that a lot of damage can be done if debriefing is handled badly, and those selected to carry out debriefing should show a degree of sensitivity.

The role of simulators in assessment

Many people think of simulation as a tool for assessment, and in particular for certification or revalidation. They believe that a session in a simulator will reveal whether that anaesthetist is competent or not. Limitations of the models, partly due to our lack of understanding of the complexity of the human being, and partly due to limitations in technology prevent us achieving the level of fidelity that would be required to make judgements of competence with confidence. We may be able to make judgements about aspects of performance, in the way that we can tell whether someone can perform external cardiac compressions properly or not, by looking at their performance on a manikin. However, if we are attempting to assess the total performance, then that lack of reality will be significant. There are other challenges to assessing performance and these relate to the lack of a valid, reliable and feasible scoring system. Byrne and Greaves recently reviewed the literature relating to in simulator assessment systems and concluded that there was no scoring system available that had been satisfactorily validated.[9]

It is certain that further work will result in more satisfactory assessment tools to use in simulators and that these will then be used in all types of assessment for anaesthetists.[10]

The strengths of simulation as a formative or diagnostic assessment tool have already been dealt with in detail.

Education research

Simulators in anaesthesia are thought of mainly in terms of their contribution to health care curricula. Their role is not confined to this field because they are also used in research. Many of the articles published in connection with simulation are concerned with the educational roles. However, work has been done on human factors issues and these have been summarised by Dave Gaba.[12]

Human factors research (Gaba)[1]

- Investigation of information gathering and decision making techniques.
- Investigation of human-machine interaction with clinical equipment.
- Development of techniques and tools for performance assessment, evaluation and testing.
- Investigation of the effect of training on 'clinical' performance.
- Investigation of the effects of 'performance shaping factors' on 'clinical' performance; for example fatigue/sleep deprivation, age, stress, noise etc.
- Testing of user interfaces of new monitoring or therapeutic devices.
- Validation testing of medical intelligence systems, including comparison with human performance on the same simulated scenarios.

Education feedback

Most of the emphasis on simulation is in its role in the educational development of individual learners. However, when trainees from the same department or school of anaesthesia undergo the same scenarios it becomes possible to get information on what aspects are done consistently well, and which aspects are not performed to the desired standard. This information can be fed back to those in charge of the training programme, so that changes can be made to that programme. We have some experience of this in the Scottish Centre, but have not yet closed the loop by seeing the effect of changes suggested by our analysis of performances.

References

1. Gaba DM. Simulators in anaesthesiology. In: Lake CL, Rice LJ, Sperry RJ, editors. *Advances in Anaesthesia, Volume 14*. St Louis, Missouri: Mosby, 1996.
2. Schwid HA, O'Donnell D. The Anesthesia Simulator Consultant: simulation plus expert system. *Anesthesiology Review*. 1993;20:185–189.
3. Smith NT, Starko K. PC-based simulators. *Society of Computing and Technology in Anaesthesia News*. 1995;8:8–9.
4. Byrne AJ, Jones JG. Responses to simulated anaesthetic emergencies by anaesthetists with different durations of clinical experience. *British Journal of Anaesthesia*. 1997; 78:553–556.
5. Good ML, Gravenstein JS. Anesthesia simulators and training devices. *International Anesthesiology Clinics*. 1989;27:161–168.
6. Gaba DM, DeAnda A. A comprehensive anesthesia simulation environment: recreating the operating room for research and training. *Anesthesiology*. 1988;69:387–394.
7. Christensen UJ, Andersen SF, Jacobsen J, Jensen PF, Ording H. The Sophus anaesthesia simulator v. 2.0. A Windows 95 control-center of a full-scale simulator. *International Journal of Clinical Monitoring & Computing*. 1997;14(1):11–16.
8. Chopra V, Engbers FH, Geerts MJ, Filet WR, Bovill JG, Spierdijk J. The Leiden anaesthesia simulator. *British Journal of Anaesthesia*. 1994;73:287–292.

9. Byrne AJ, Greaves JD. Assessment instruments used during anaesthetic simulation: review of published studies. *British Journal of Anaesthesia.* 2001;86:445–450.
10. Glavin RJ, Maran NJ. Development and use of scoring systems for assessment of clinical competence (Editorial). *British Journal of Anaesthesia.* 2002;88:229–330.
11. Gaba DM, Howard SK. Situational awareness in anaesthesiology. *Human Factors.* 1995;5:20–31.

Further reading

Gaba DM, Fish KJ, Howard SK. *Crisis management in anesthesiology.* New York: Churchill Livingstone, 1994.

19

Simulation and technical skills

Ronnie Glavin and Nicki Maran

In this chapter we shall review the role of simulation in teaching and assessment of technical skills. By technical skills we mean the use of medical expertise, drugs and equipment; the theoretical knowledge underpinning anaesthesia and its application along with the use of practical skills.

Technical skills range from the simple to the complex, from specific facts and techniques to integrated performance. Not only is there this range of technical skills but there is also a difference in the way in which learners acquire skills and deal with them. The Dreyfus Brothers' model gives us characteristics for the different level of learner (Table 1).[1]

Table 1.

Dreyfus Brothers model of	Skills acquisition
Level 1 Novice	Rigid adherence to taught rules or plans. Little situational perception. No discretionary judgement.
Level 2 Advanced beginner	Guidelines for action based on attributes or aspects (aspects are global characteristics of situations recognisable only after some prior experience). Situational perception still limited. All attributes and aspects are treated separately and given equal importance.
Level 3 Competent	Coping with crowdedness. Now sees actions at least partially in terms of longer term goals. Conscious deliberate planning.
Level 4 Proficient	Sees situations holistically rather than in terms of aspects. Sees what is most important in a situation.

We can make use of these differences when designing teaching packages for the different levels of learner.

Simulation has been used for a long time to develop practical skills and the advantages are listed by Holzman.[2]

Advantages of simulators for teaching technical skills.[2]

- Undesired interference can be reduced or eliminated.
- Tasks are relevant.
- Skills can be practised repeatedly.
- Retention and accuracy are increased.
- Risk to patients and to learners is avoided.
- Suitable clinical cases are always available.
- Standards against which to evaluate student performance and diagnose educational needs are provided.
- Transfer of training from classroom to real situation is enhanced.

It is useful to divide technical skills into two groups for the purpose of planning teaching.

- *Normal practice*: the management of the type of clinical challenge we would expect to meet on a regular basis e.g. patients with full stomachs requiring emergency surgery.
- *Unusual cases*: those we might not see very often throughout the course of our professional practice but cases that we would be expected to manage successfully e.g. anaphylaxis.

Teaching normal practice

Teaching novices

Novices are characterised by rigid adherence to strong rules. They understand a right and a wrong way to do something. Novices need rules. If we as teachers do not provide rules, then they will adopt rules of their own. It therefore makes sense to provide them with rules. Specifically we should provide them with the rules that we want them to apply and to use at this stage of their anaesthetic training. Many anaesthetic departments already provide rules and drills for novices to learn. How can simulation play a

The learning points in connection with learning a technical rule are:

- When to apply the rule.
- When the rule should not be used (contra-indications to the drill).
- The ability to perform the drill to a recognised standard.

role? We can create scenarios that will give novices the opportunity to apply a rule and to execute the drill associated with that rule. We can relate the advantages listed by Holzman to this situation.

- *Undesired interference can be reduced or eliminated.* We can create a scenario where we are only interested in the rules and the drill. It may not be easy to achieve this in a real patient. It may also be difficult to review the performance of this drill in a real patient, because many tasks that require our attention throughout the surgery.
- *Tasks are relevant.* The scenario deals only on the features of the anaesthetic relevant to the particular rule and drill we wish to promote.
- *Skills can be practised repeatedly.* We can stop the scenario as required, and either repeat the same case or move on to another case. This is seldom possible in the real world.
- *Retention and accuracy are increased.* By practising repeatedly we can develop the drill, so that it becomes more automatic. The ability to repeat the drill immediately gives the opportunity to correct any inaccuracies. If the novice doesn't do things in the correct sequence, or forgets one stage, then he or she can repeat and attempt to improve on the performance of the previous effort.
- *Risk to patients and to learners is avoided.* It is important for the learner's sense of professional development that he or she can recognise that progress is occurring. Failure to carry out the drill on a real patient may have an adverse effect on the learner's confidence, which may put more pressure on the learner for the next occasion on which the drill is to be performed. The lack of risk to the patient is also consistent with promotion of longer-term goals, such as commitment to patient safety. It is easier to promote this value by being seen to observe it as a teacher, and insisting that the drill is practised to the necessary standard on a simulator than to encourage a novice to try out a new drill or technique on a patient.
- *Suitable cases for teaching can be simulated as required.* We can use simulation to provide a trainee with a variety of indications to use the drill. We can provide a range of scenarios with these indications. For example, in the case of rapid sequence induction, we can have a patient with a full stomach due to a recent food or drink intake, we can have a patient with a full stomach due to delayed gastric emptying, such as with administration of parenteral opioid drugs for pain relief, we can have a patient with significant gastro-oesophageal reflux etc. Similarly we can explore relative contra-indications to the rule e.g. a predicted difficult intubation due to limited mouth opening and fixed neck flexion in a patient with severe ankylosing spondylitis.
- *Standards against which to evaluate student performance and diagnose educational needs are provided.* The standard can be set by the teacher demonstrating performance of the drill in real time. This technique is used in Advanced Trauma Life Support and Advanced Life Support courses and it provides the learner with

a mental image of how the drill should be performed. A useful approach is to have a group of learners critique the performance of each other by using a checklist. It is also possible to have the novice critique his or her own performance by video-recording the performance, and asking that learner to use a checklist against the recording. This has the advantage that those areas identified by the learner are more likely to be acted on than areas identified by the teacher.
- *Transfer of training from classroom to real situation is enhanced.* A common observation of those of us working with simulators is that learners can often provide correct written answers, but may not be able to translate that knowledge into effective action. In educational terms they can recall information (the lowest cognitive level) but may not be able to successfully apply that information (two cognitive levels higher). During clinical scenarios the novices have to demonstrate that they can seek out the relevant information from the patient, case records etc. to determine whether indications or contra-indications are present. This provides the opportunity for them to review not only their knowledge but also their clinical skills. Learning in a context that resembles the one in which the knowledge will be applied is also thought to improve that learning. Learning in an environment that is more like the workplace than a classroom should therefore help make the training more effective.

When designing the scenario we need to ask ourselves some questions. What educational points are we trying to promote? What will constitute a satisfactory performance? Under what conditions do we expect the trainee to perform? What knowledge and skills do we expect the learners to possess before coming to the training exercise?

The conditions under which the drill is to be performed can refer to the clinical condition of the patient, and so include indications and contra-indications. However, we may also set some rules as to the level of supervision under which the drill can be carried out. This is of relevance in the UK, where the consultant or attending anaesthetists need not be physically present when residents or trainees are managing some emergency cases.

We can also specify the knowledge and abilities that we would expect trainees to possess before undertaking this type of training. In controlled, rapid sequence induction we would expect that trainees are able to perform endotracheal intubation to a satisfactory standard, and have some knowledge of the pharmacology of the common intravenous induction agents and muscle relaxants.

What kind of simulator should be used for novice training?

The educational goals are that the novice should be able to demonstrate the ability to apply the rule and should be capable of executing the drill.

The scenario can therefore be divided into two stages. The first stage is eliciting the relevant clinical information, and the second stage is carrying out the drill. The first stage can be achieved by having a faculty member play the role of the patient (or a faculty member briefing other novices in the group). This can be supported by using a 'dummy' set of case notes with further information. The main components for the second part are a faculty member playing the part of the anaesthetic assistant, and the availability of associated items of equipment such as suction apparatus, a tilting trolley, laryngoscopes, dummy drugs etc. In the case of the rapid sequence induction that we have discussed previously, the interaction with the manikin is confined to intubation, so the level of sophistication of the actual manikin need not be high for novice training. Indeed placing an intubation part task trainer in the operating theatre may be of more use to the novice for this particular task, than taking the trainee to a simulation centre.

Assessment of the novice anaesthetist

Simulation can be used for the formal assessment of the drills with which novices need to be familiar before they move to practice on real patients. In the UK, the ability to perform a controlled rapid sequence induction should be assessed at about three months into anaesthetic training. Simulation allows such an assessment to be carried out at a time that is mutually convenient to both trainee and assessor. Although success in this area alone does not guarantee progress to the next phase of training, it should help improve the trainee's confidence in his or her ability to carry out the technique. Any components of the technique that are not performed to the required standard can be addressed immediately, and the test can be repeated as soon as both parties desire.

The question of where the standards for performing a drill come from is too large to discuss at this point, but as a general principle we can use published standards if available. If none are available then we can set standards that will at least meet with the approval of our colleagues. We do not want novices being criticised for doing something wrong, when in their mind they are applying the rule and performing the drill to the standard they have been given in the teaching session.

Teaching the advanced beginner

Experience is now beginning to influence the way in which the learner acts. The rigidity associated with the novice level is decreasing. We can now modify the major rules by introducing subsets of rules. We can also encourage the learner to explore the boundaries of the rules. In the previous section the emphasis was on keeping things simple: teach one way to perform a controlled rapid sequence induction and give rules as to when this technique should be used and when it should not be used. Now that the learners can cope with some subtlety we can introduce some variation. Let us continue with the example used in the previous section — controlled rapid sequence intravenous induction of anaesthesia. Other factors can now be introduced that will modify the rule and explore alternative methods of carrying out the drill.

Examples of alternative ways of carrying out the drills would be using techniques that avoid suxamethonium, and achieving conditions for intubation by using propofol and a short acting opioid.

To design suitable lessons identify the learning points, conditions, pre-existing knowledge, practical skills and experience that are expected and present the learners with a simulation that calls for their use.

Scenario 1 — Background information.

You are the on call anaesthetic registrar in a medium sized hospital.
You have been asked to see a 21-year-old male farm hand who has sustained a compound fracture of the right tibia, with contamination of the wound, having been kicked by a horse. The orthopaedic surgeons wish to take him to theatre as soon as possible to debride the wound and reduce the fracture.
The patient has eaten a cheeseburger, fries and milkshake one hour prior to the accident. He has received pethidine 100 mg intramuscularly in the accident and emergency department 20 minutes ago.
The patient has asthma (since childhood) which is normally well controlled but he has been a little wheezy since the incident. The patient does not wish a regional block.

Course.

Trainee is presented with patient details.
Patient's wheeze will improve with use of inhaled beta 2 agonists.
Pressure is put on trainee by orthopaedic surgeon (faculty member).
Senior anaesthetist (faculty member) is involved in an emergency case elsewhere but is able to discuss details and is happy for the trainee to proceed with the case.
If suxamethonium is used in the controlled rapid sequence intravenous induction then the patient will develop bronchospasm.
This will show as increasing rise in Peak Airway Pressure, decrease in expired tidal volume and a decrease in venous return resulting in increase in heart rate and fall in blood pressure. These changes will increase in severity until treatment begins.
Bronchospasm will respond to simple measures such as deepening the level of anaesthesia and administering beta 2 agonist drugs.

Scenario 1 — Learning points (continued).

The main learning point is to explore the risk/benefit ratio of using suxamethonium by dealing with a relative contra-indication.
Suxamethonium can cause histamine release and may not be a good choice in a patient with asthma whose asthma is not controlled.
Patients with irritable airways — heavy smokers, chronic bronchitis sufferers — are also more likely to respond to the stimulus of intubation by developing bronchospasm.
Alternative methods of performing controlled rapid sequence intravenous induction are available and may be indicated in such a case.
Methods of obtunding the response of the trachea to stimulation, and methods of preventing the trachea subsequently becoming irritated by the endotracheal tube can also be considered and discussed.

The decision to make the bronchospasm respond to simple measures was taken because the learning points do not include the management of bronchospasm or the management of unstable asthma. However, the bronchospasm should result in significant changes in the respiratory parameters and cardiovascular parameters, to show that there is a potential risk attached to the administration of suxamethonium in these circumstances. Management of intractable bronchospasm or unstable asthma can be covered in debriefing, but should take second place to the main learning points.

In this scenario the patient's condition illustrates a relative contra-indication of suxamethonium. It would be wrong to give this scenario to a novice because the wrong message may be taken on board. An inappropriate but 'strong' rule may be introduced — 'Do not use suxamethonium in asthmatic patients'. As stated in the intended learning outcomes part the scenario is to encourage trainees to explore the relative contra-indications of suxamethonium in emergency cases, and to gain some experience of using alternative drugs.

What kind of simulator(s) can be used for this scenario?

As with the scenarios at the novice stage the role of the patient can be played by a member of faculty. However, once anaesthesia is underway the scenario should have a dynamic component in that the changes will worsen until treatment starts.

Although it is possible to use a basic part-task intubation trainer connected to an anaesthetic machine, the trainee will then have to rely heavily on a faculty member to provide information. This may be acceptable in this scenario because the main learning points are not the detection of the changes, but the impact on the patient of this potential side effect of suxamethonium.

Assessment of advanced beginners

The broad principles are similar to the novice section. However, by this stage in their anaesthetic training the differences in clinical experiences, combined with differences in the rate at which learners acquire knowledge, will probably result in greater variation in knowledge and skills than with novices. The formative role becomes very important in helping not only to identify gaps, but also to help each learner compare his or her knowledge with that of the other members of the group.

Teaching the competent trainee

The learner is now looking at the bigger picture and is able to consider the longer-term goals. So, rather than seeing the management of the patient in terms of a series of sequential tasks, the learner is looking at the management in more global terms. In addition, the nature of clinical practice, with a large and variable range of patient features, makes it difficult to apply exact rules, or work in an algorithmic manner. We are now into the realm of heuristics, or rules of thumb. The anaesthetist is acting more like an analogue computer where all the information is integrated to form a larger pattern that will influence how the anaesthetist tackles the challenges.

What then are we trying to teach at this level in terms of technical skills? We expect the learner to be obtaining knowledge from clinical experience supplemented by reading textbooks and journals, attending meetings etc. Many practical skills will have been learned and refined in clinical practice. There may be some skills that have yet to be mastered, such as awake fibre-optic intubation, but in general by this stage we are more concerned with the successful integration of individual competencies into an overall performance. Can the learner use the technical skills already acquired to successfully manage more challenging clinical cases?

Scenario 2 — Background information.

You are the on-call anaesthetic registrar for a medium sized hospital.
You have been called to A&E department to help manage a 25-year-old cyclist involved in collision with car. Patient has sustained a head injury (GCS falling from 15 when first seen by the paramedics to a present score of 8) and a fractured left femur. He has a receding chin and prominent upper incisors. A friend of the patient confirms that he last ate about one hour prior.

Clinical course.

Patient will have signs consistent with hypovolaemia and falling GCS.
Unless the hypovolaemia is treated it will worsen.
Failure to maintain a satisfactory PaO_2 and $PaCO_2$ will result in a rise in intracranial pressure and further deterioration of GCS score.
Failure to obtund the stress response to intubation will also result in a rise in intracranial pressure and further deterioration of GCS score.

Scenario 2 — Learning outcomes (continued).

Recognition of the following potential problems: hypovolaemia, full stomach, head injury and difficult intubation.
Management of this patient in such a way that he will not become more unstable as a result of anaesthetic interventions.

There is no one anaesthetic technique that can be regarded as the only way in which this patient can be managed. It is more important to follow good principles, such as those described in the learning points and clinical course. For example, the trainee may manage hypovolaemia by intravenous fluid therapy and may choose to use a smaller dose on intravenous induction agent if inducing anaesthesia that way. However, a smaller dose of IV induction agent may have implications for the stress response to intubation — which may be worse because of prolonged attempt at intubation from the difficult conditions (anatomical problem plus cervical collar). In this type of scenario the rules cannot be applied in a rigid manner, because of potential conflict. The trainee has to modify rules in pursuit of principles of management. Algorithms have only a limited place here because their success cannot be guaranteed by moving through the problems in a sequential manner. This is clearly moving away from the application of hard and fast rules. There are a series of conflicting goals that must be assessed and weighed up. High fidelity simulators are very effective in providing these kind of scenarios, where many options are available to the participant, and rather than attempt to reward one 'proper' solution the script will punish dangerous actions. For example, in the above scenario, allowing the patient to become hypoxic and hypercarbic could worsen the head injury and brain swelling, so causing intracranial pressure to rise with adverse effects. How the learner avoids these conditions is his or her choice. The second part of the educational exercise is the review of the performance. Donald Schon[3] refers to this type of 'reflection-on-action' as a way of encouraging professionals to try and understand their actions so that they can learn from them for future application.

Assessment of the competent trainee

At this level, with no one correct answer, it becomes more difficult to use simulation as a tool of summative assessment, not least because the criteria for successful management become more difficult to define. It offers advantages in formative assessment, because the process of review can help identify strengths and weaknesses and help the learner sort out the relative priorities.

Teaching at the stage of proficiency and expertise

The next stage is to integrate the non-technical skills along with the technical skills. This will be dealt with in greater detail in the next chapter.

Teaching unusual conditions

Many anaesthetists equate simulation with unusual conditions. The reasons are obvious. By their very name these are conditions that very few anaesthetists will encounter during their clinical practice. Yet, when they do occur, prompt recognition and appropriate management is necessary, if the patient is to remain free from harm. We can divide simulator training into these two phases — recognition and management. We also have to take into account the stage of training of the learner. At novice level the rules must be strong and unambiguous, at advanced beginner level we can introduce some variation, and by the level of competence we can explore variations in greater depth. In this section we shall use anaphylaxis as an example of an unusual condition.

One of the difficulties with simulation is that learners are much more alert for abnormal conditions and things going wrong than would be the case with real practice. One of the dangers of anaesthesia is its current level of safety, where we do not expect things to go wrong. We shall have cases that are challenging because of the surgical condition and procedure, accompanied by the medical condition of the patient — a patient with severe ischaemic heart disease and arrhythmias requiring repair of a leaking aortic aneurysm will always be challenging, but we can anticipate the likely problems. The danger with the unexpected conditions is their appearance when we are not expecting things to go wrong.

At novice level in anaesthesia we are dealing with doctors who in the course of their pre-anaesthetic training should be familiar with aspects of anaphylaxis. They should know the clinical pattern and should know the basic treatment principles. There are several rules that are being promoted in this type of scenario at this level. First of all there are the general rules — Is the patient well? If not apply ABC (airway, breathing and circulation) and call for more senior help. Then there are the specific rules — if clinical pattern 'X' is found then apply treatment protocol 'Y'. Trainees are much more sensitive to cues on a simulator than in the real world and will sometimes 'detect' anaphylaxis when it is not actually taking place. Novices do not have the ability to see the totality of the case and are much more likely to react to one specific cue.

Scenario 3 — Background information.

A 23-year-old female is to undergo an emergency appendicectomy. General health is good. The patient is stable — no signs of systemic sepsis. No electrolyte imbalance noted, vital signs consistent with disease.
Immediately after intubation the patient's blood pressure falls, heart rate speeds up and airway pressure rises. Faculty member, playing part of anaesthetic assistant, observes that the patient has a widespread rash.

Scenario 3 — Clinical course and learning outcomes (continued).

Clinical course.

The patient will respond to epinephrine 0.5 mg intravenously, chlorpheniramine 10 mg iv and hydrocortisone 200 mg iv plus iv fluids. Secondary treatment of bronchospasm will not be required at this stage.

Learning outcomes.

The major learning outcome is the recognition and early management of anaphylaxis.
The recognition can be made more difficult by having no skin changes.

The scenario scripted as in the box avoids complexity and ambiguity. The onset after intubation makes the management easier, because the patient will already have an endotracheal tube in place. The challenges of intubating a patient with severe upper airway obstruction due to swelling of the vocal cords and supraglottic structures are omitted. At this stage the learner is only being asked to recognise and treat the basic condition.

At advanced beginner level the scenario can be made more complex by making the diagnosis more difficult e.g. by only having cardiovascular changes, or by making the management more difficult e.g. by having bronchospasm persist following the initial dose of epinephrine.

Scenario 4 — Background information.

During the management of an otherwise well patient undergoing hysterectomy for large uterine fibroids there is a sudden blood loss of 500 ml. Following the administration of intravenous fluids the patient's blood pressure falls and heart rate increases. There are no early skin changes nor does the airway pressure change initially.

Clinical course.

The surgeon is confident that he/she has secured haemostasis. The patient's condition will continue to deteriorate until the anaesthetist administers epinephrine 0.5 mg intravenously, chlorpheniramine 10 mg iv and hydrocortisone 200 mg iv plus iv fluids. A further dose may be required if the anaesthetist does not detect the condition quickly. 5 minutes after the onset of cardiovascular changes the airway pressure will begin to rise.

Learning outcomes.

The major learning outcome is the diagnosis and management of anaphylaxis but within a clinical context where there are many variables changing. The changes in respiratory system and in the skin are intentionally left until later so that the diagnosis is not easy. This reflects the challenges that can arise in diagnosing this condition. The underlying trigger agent may not be obvious at this stage.

When we reach the level of competence then we can make the total picture more complicated by placing the anaphylactic episode in an already dynamic situation.

Our experience with these scenarios is that most trainees will work through drills reasonably well, but are often slow to diagnose the condition. Even experienced anaesthetists can take some time to commit themselves to the diagnosis. This probably reflects unfamiliarity and the consequences of committing the patient to that course of treatment. This leads us into discussion about cue recognition, pattern matching and making use of information. These we shall deal with in the next chapter.

References

1. Dreyfus HL, Dreyfus SE. *Mind over machine: the power of human intuition and expertise in the era of the computer*. 1986; Oxford: Basil Blackwell.
2. Holzman GB. Clinical simulation. In: Cox KR, Ewan CE, editors. *The medical teacher*. London: Churchill Livingstone, 1988;240–243.
3. Schon DA. *Educating the reflective practitioner*. San Francisco, California: Josey Bass, 1987.

20

Simulation and non-technical skills

Ronnie Glavin and Nicki Maran

Anaesthetists' non-technical skills:

- Situation awareness.
- Decision making.
- Task management.
- Team working.

The conventional model of teaching in medicine places great emphasis on the acquisition of knowledge and practical skills. However, possession of these alone by a doctor does not guarantee successful patient management. Something else is required, and those abilities are referred to collectively as non-technical skills. In this chapter the role of simulation in teaching and assessing non-technical skills will be explored.

Aviation led the way in this field.[1] Following some high profile disasters in the 1970s, researchers in the field concluded that improving 'stick and rudder' skills did not appear to be having an impact on safety. What seemed to be missing from flight crew involved in these high profile disasters was a lack of other skills — non-technical skills. These skills included the ability of the individual crewmembers to process information, make effective decisions and work effectively as a team. Other industries where safety is given high priority, and where people work together in dynamic situations with conflicting goals and information never fully available, have adopted this approach, now known as Crew Resource Management. In medicine, anaesthesia has led the way with a course adapted from an aviation model by Professor David Gaba at Stanford, California. This course, Anesthesia Crisis Resource Management (ACRM),[2] is now taught in many countries.

In this chapter the application of simulation to the principles of teaching and assessing these non-technical skills will be reviewed. The first challenge is to identify the non-technical skills necessary for anaesthetists. The system described by Fletcher in an earlier chapter, the taxonomy of anaesthetists' non-technical skills and associated behavioural markers, will be used in this chapter.

Good, effective clinical performance requires that all the skills — technical and non-technical be integrated. However, we shall look at each of the four main categories individually, with the understanding that this is a step towards integrated performance.

Situation awareness

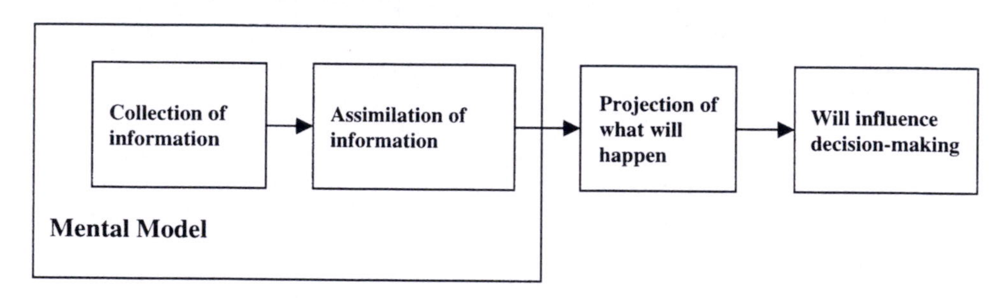

Figure 1.

The key components of situation awareness are shown in Fig. 1. Information is collected and then processed. These two stages allow the formation of a mental model which allows projection into the future of possible courses of events. Projection will in turn influence the decisions taken by the anaesthetist.

Information gathering is an active process. A lot of information is available to us in the operating theatre, and we have to actively seek out the particular information that will help us manage the patient effectively. It is difficult for humans to cope with lots of separate pieces of information, so we make this process easier by combining the individual pieces of information into collective patterns.

For example, a patient with a weak peripheral pulse, a low end tidal carbon dioxide, a tachycardia, normal ECG complexes, and a low blood pressure is likely to have a reduced cardiac output, possibly from hypovolaemia. If we also notice a marked and sudden fall in saturation then we may think of other conditions, such as tension pneumothorax. Part of learning to become an anaesthetist is to learn how to handle information — what to look for and when to look for it. For example, immediately after intravenous induction we look at pulse rate and blood pressure, whereas after an attempt at endotracheal intubation we look at and listen to the patient's chest and look at the capnograph. As one becomes more experienced this process becomes more automatic. Although these skills are acquired during the management of real patients, simulation may have a complementary role to play in the development of this process.

Simulation can help to demonstrate these patterns by showing the way in which the variables change, as the underlying condition develops. For example, review of video-recording of a scenario in which there is sudden, unexpected blood loss due to a surgical complication can be used to draw the trainee's attention to the changes in the different parameters as the blood loss mounts. The dynamic pattern can be studied in the context of the changes taking place in the patient.

Fixation errors

- *This and only this* — The anaesthetist becomes fixated on one piece of information to the exclusion of all other information.
- *Anything but this* — In this case the anaesthetist refuses to commit to a particular course of action despite strong evidence that it is there.
- *Everything is O.K.* — This is denial that anything is wrong despite clear indications of a problem.

Effective information gathering is necessary if the mental model created by the learner is to accurately reflect reality. For example, in a case of hypovolaemia due to blood loss one would expect to see an increase in heart rate, a drop in blood pressure and a drop in end tidal CO_2. If the patient also developed wheezy breath sounds and an increase in airway pressure during the inspiratory phase of ventilation the cause could be an anaphylactic reaction, and not blood loss. On the other hand, if a trainee only detected an increase in heart rate, and ignored, or did not detect the other changes, then an inaccurate mental model may be developed. If the trainee decided that the increase in heart rate was due to 'light' anaesthesia and consequently increased the inspired concentration of volatile agent, then an adverse outcome for that patient could ensue. Persisting with an inaccurate mental model, especially when evidence to the contrary is available, is described as a fixation error. Gaba[2] describes three of these; 'this and only this', 'anything but this' and 'everything is OK'. These do occur in clinical practice but their incidence will vary. They can be encouraged in the simulated clinical environment.

Scenario 1.

An 18-year-old male is admitted for an emergency appendicectomy.
The patient is generally fit apart from requiring a beta 2 agonist inhaler for exercise related 'tightness' in breathing.

Immediately following induction and connection to the breathing system the patient's airway pressure is high and over a 5–10 minute period his oxygen saturation falls to around 85%. There is no wheeze on auscultation of chest. The cause is a foreign body in the connection to the breathing circuit.

In a review of the first forty nine times we ran this scenario we observed all three types of fixation error. Many trainees committed a 'this and only this' error by ascribing the high airway pressure to bronchospasm and proceeded down the path of management of bronchospasm. Many also committed an 'anything but this' fixation error, by refusing to consider possible equipment causes, despite this possibility being suggested by a member of faculty. Fortunately, only a very small number committed the third type of fixation error 'everything is OK', by carrying on as though there were no problem and allowing the surgeon to proceed with the operation.

Encouraging trainees to identify these fixation errors during the review of the scenario is only half the exercise. Trainees must also be given advice and instruction on skills and strategies to avoid such an error occurring.

This sort of training is often associated with simulation. Indeed Gaba's ACRM course, which includes these areas, is almost synonymous with anaesthetic simulation, which is not surprising, because he developed the simulated Complete Anaesthetic Environment to study and teach some of these non-technical skills. However, low fidelity simulations, such as written description of cases and pen and paper exercises, can also be used in promoting non-technical skills. Expensive simulators are therefore not essential in teaching non-technical skills, but certainly give excellent opportunities for trainees to apply their newly acquired non-technical skills knowledge and abilities.

Scenario 1 was useful in demonstrating the lack of an accurate mental model in the anaesthetist. Effective patient management requires that *all theatre staff* share an accurate mental model of what is going on. Mental models should not therefore be thought of as applying to only the anaesthetist. Techniques for sharing mental models and maintaining an accurate mental model for the group must be included in these types of courses. Simulating the clinical environment allows practice in such techniques, but lower fidelity models can also be used.

Accurate mental models allow projection which influences decision making — this leads naturally on to the next section.

Decision making

There is a spectrum of decision making. At one end options are consciously listed and the relative pros and cons calculated, while at the other end the process is so intuitive that the person making the decisions may not even be aware of doing so. Intuitive decision making develops with experience, and is therefore doesn't play a key feature in decisions made by novices or advanced beginners. One way in which simulation may help, is by providing a wider range of experience for the learner, and so help to expedite the transition from advanced beginner to competent practitioner, as described by the Dreyfus Brothers model in the previous chapter. However, just doing cases may not be enough. It is recognised in aviation that flight experience does not equate to expertise. Jensen[3] lists factors relating to personal experience that help make this transition in aviation decision-making. Simulation can help increase the variety of experiences, and reflective debriefing can help increase the meaningfulness of the experience.

Aspects of personal experience that help develop expertise:
- Number of experiences.
- Variety of experiences.
- Meaningfulness of experience.
- Relevance of experience.
- How recent the experience was.

The major challenge is to help trainee anaesthetists make better decisions. How can they improve their clinical decision making abilities? Good decisions require that options are generated, evaluated and the most appropriate one then selected. The process does not end there, because the effects of the actions on the patient, are reviewed — which may modify the decision. This latter step is very important in the management of unusual conditions. In Scenario 3 in the previous chapter, the first decision to make is deciding what has happened. Options are generated. Could it be septicaemia, hypovolaemia, anaphylaxis etc? The relative advantages and disadvantages of treatment are considered. Are there likely to be harmful effects from epinephrine in this patient? When epinephrine, chlorpheniramine and hydrocortisone, oxygen and fluids have been given the effect of those on the patient has to be reviewed. Is the patient improving? Was this the correct diagnosis? The process is iterative because the provisional diagnosis will prompt actions and review of those actions can help confirm or refute that diagnosis.

It may seem obvious, but for trainee anaesthetists to learn to make effective decisions they have to be allowed to make decisions in the first place, and then follow through the consequences of their decisions. This is becoming more difficult in the UK. Anaesthesia trainees work under closer levels of supervision than previously. Simulation can help by putting learners into positions where they have to make decisions and live with the consequences of those decisions. Making decisions is vital, but it is not the only part of the process. Those decisions must be reviewed to determine whether good or bad choices were made. Donald Schon[4] refers to this process as reflection on action. As mentioned in previous chapters on simulation, this can be difficult to conduct in the real world. At the simulator the opportunity to see events as they unfold, undistorted by memory, can help this process. For example, in the anaphylaxis scenario (Chapter 19, Scenario 3), the trainee can review the options generated, the analysis of those options, the timing of intervention, the appropriateness of intervention and the effects of intervention. The importance of reflection is that it is not confined to simulation centre, but that it can be developed during simulator courses and subsequently used in clinical practice.

Task management

This is described as 'The effective management of resources and organisation of tasks to achieve goals'. This breaks down into two obvious phases. The first phase is determining the goal. This may be the peri-operative management of the patient. The second phase is then deciding what resources are required. Resources can be summarised as information, personnel and equipment.

In reality, of course, the boundaries are blurred. Achieving the goal will only be possible when one has some information about the patient and the nature of the surgical procedure.

Scenario 2.

A 55-year-old man is scheduled to undergo elective abdomino-perineal resection for adenocarcinoma of rectum. The patient has had two myocardial infarctions, the most recent being 5 months ago and claims that he was informed that he was difficult to intubate.

If we consider the case in Scenario 2 then we clearly need some more information about the patient. What is his current cardiac function? What was required to manage his 'difficult intubation'? The answers to these questions and all the usual preoperative questions will determine how this patient is to be managed — awake fibre optic intubation, pulmonary artery flotation catheter inserted under local preoperatively and the patient's condition optimised, the use of regional analgesia?

Once the decisions have been made then the necessary resources have to be in place. This is standard anaesthetic practice and can be taught in the real world on real patients. However, these skills can be developed using simulation. Paper and pen exercises such as described for Scenario 2 can be used to help establish the principles. Using simulated patients (member of faculty playing patient) can likewise encourage the appropriate form of questioning and discussion of further investigations and possible plans for management.

As all anaesthetists know, no matter how well prepared we think we are, circumstances can change and plans may then have to be reviewed.

Scenario 3 describes a patient whose management presents the anaesthetist with some challenges. If the patient's testis is to be saved, then surgery should not be delayed. On the other hand, there are the potential risks associated with the full stomach.

The trainee 'hot-seating' this scenario has to recognise that Malignant Hyperpyrexia is taking place and then has to formulate a plan to deal with that. MH has been chosen because there are many components to be dealt with for the task to be successfully managed. These include changing the breathing system, arranging intravenous anaesthesia, mixing dantrolene, getting ice, establishing invasive monitoring, arranging biochemistry and haematology investigations etc. Prioritisation of tasks, communication with other personnel and making effective use of the other theatre personnel will all feature in effective task management. Setting up a scenario like this, in the simulated theatre environment, will give participants the opportunity of applying the skills of task management. This can be aided by reviewing the performance and commenting on how the task management was effectively realised.

Scenario 3 — Background information.

A 19-year-old male presents with an acute torsion of the testis of two hours duration. The surgeons wish to operate immediately. The patient has had burger, large fries and a milk shake about 1 hour prior to onset of torsion. His general health is otherwise good.

Clinical course.

The patient does not wish a regional technique. Induction and intubation are uneventful. Following preparation for surgery the patient's heart rate rises, his blood pressure becomes unstable, his end tidal CO_2 rises and he develops arrhythmias.
This will respond to intravenous dantrolene 2 mg/kg, stopping the trigger agents and cooling the patient. If intervention is not prompt the patient's serum potassium level will rise and the arrhythmias will become life threatening (VT with a pulse progressing to VT without a pulse progressing to VF).

Learning outcomes.

Recognition of signs of MH.
Management of Malignant Hyperpyrexia, which will include the effective use of resources.

Team working

This is defined as 'the skill of working with others in a team context, in any role, to ensure effective joint task completion and team satisfaction'. The emphasis is now on the team rather than the task.[5]

In anaesthesia we work with many health care professionals in the delivery of patient care, and it is a matter of debate as to whether we work as part of one large team or whether we fulfil roles in many small teams. What then can we do in simulation to improve the ability of learners to work as part of a team?

Familiarisation with the role of other team members

If a task, such as effective patient care, is to be administered effectively, then each member of the team should be making contributions to that goal. A basic question is not only what does each person do, but what can each person do. This becomes of increasing importance during emergencies or crises, when we may have to call upon resources not normally used during routine practice. For example, during normal routine practice in the operating theatre, on an elective list, we expect the nursing staff to be able to help the surgeon with instruments, prepare equipment for the next case and organise the smooth flow of patients. Events such as in Scenario 3 require that the non-anaesthetic members of the operating team depart from their 'normal' routines and take on other duties. In some cases this may require skills that are not normally displayed. If the patient developed pulseless VT secondary to hyperkalaemia then we would expect team members to perform chest compressions on the patient while defibrillation was being organised, even though this activity would not be part of normal daily practice.

Allocating tasks to the most appropriate team members is only possible when one knows what each team member is supposed to be able to do and can actually do.

Advantages and disadvantages of participants "Playing Out" their role

Advantages and disadvantages of participants playing out of role.

Advantages:	**Disadvantages:**
Allows non-hot seaters to appreciate the development of a scenario from the perspective of another role.	It is difficult for people to play roles that are foreign to them without reverting to their true role.
Keeps other participants involved during a scenario.	There is less realism when people playing out of role do not adhere to their role.
Gives some insight into the roles performed by other team members. Highlights importance of sharing mental models.	The lessons regarding team-working are not disseminated to all members of the team.

One of the advantages of having anaesthetists play the role of the scrub nurse or the circulating nurse is that it can give an impression of the development of a scenario from the perspective of that role. Many of the anaesthetists playing those parts in simulated scenarios have spontaneously commented on how little they realised that things were not well with the patient. This is partly because they are busy with their own tasks, but also because they did not have access to the same information that the anaesthetist had. This has brought home the need to share information with the rest of the theatre staff at an earlier stage if an accurate shared mental model is to be achieved. A shared mental model is more likely to result in a shared set of goals.

Define the competencies for specific tasks at both team and individual level

Within the context of the operating theatre each member of the team is there for a purpose. The surgeon is operating, the scrub nurse assisting the surgeon with instruments etc, the floor nurse fetching additional items of equipment and so on. During normal practice most members of the team will be working toward the common goal of successful execution of the surgical procedure. Shared mental models are the norm, because the staff know what is happening and what is likely to happen. During emergencies, this changes. There is less likely to be a common mental model of both what is going on and what is required to successfully resolve the problem. For example, in Scenario 3 it is extremely unlikely that any members of the theatre team will have met malignant hyperpyrexia, and so may be unfamiliar with the protocol, and therefore not know what to do. Promoting non-technical skills is difficult when the technical skills are missing, so an important component of scenarios in simulation is ensuring that participants are familiar with the technical components. Operating theatres are staffed for the normal activities of that theatre. Consequently the mixture of staff will reflect competencies suited for those activities, and applying a rigid set of tasks for each member of the team for the management of an emergency may be counterproductive. Delegation and distribution of tasks will depend on the known abilities and competencies of the individual members. Successful management of emergencies requires that the overall task (the common goal) is articulated and that each individual is aware of his or her individual duties, and where those duties fit into the bigger scheme of things.

The emphasis in educational goals will vary, depending on whether a scenario like Scenario 3 is being run for an anaesthetist, with the other roles being played by other anaesthetists, or whether representatives from the different members of the theatre team are present. If it is only anaesthetists, then the emphasis will be on setting the common goal, distributing the necessary tasks and delegating those tasks. If a whole theatre team is present, then there is more likely to be a greater need to revise some of the technical skills and to discuss the competencies of each individual member. In the former we can concentrate on the leadership role whereas in the latter we must also deal with the follower roles.

Fire drills

The above simulations are carried out in a simulation centre. This carries the following advantages.

- Review of technical skills — knowledge and practical procedures.
- Health care workers from different backgrounds can come together and work together in a scenario.
- Review facilities are more likely to be available.

There are some disadvantages also.

- It may be difficult to 'free' the required number of health care workers for a session from clinical duties.
- People are working in an environment that may be quite different from their own familiar clinical environment with different equipment etc.

If we wish to review how well the different members of different teams work together, then we may miss some things in a simulated environment. The best way of studying this is to simulate a case in the real environment. There are considerable logistical challenges, but as a review of the process, it provides the opportunity to see how well protocols are carried out. As with scenarios in the simulated clinical environment, staff must be given the opportunity to review performance so that they have involvement in the process and so will be more likely to suggest and respond to changes in practice.

The choice of whether to run a scenario in a simulation centre, or run it as a 'fire drill' will depend on the educational objectives. Simulation centre run episodes are probably better at improving the competencies, both technical and non-technical of the individual participants. Fire drills are more suited to changing practices within an organisation. There is overlap between the two and if, for example, performance of some members of the team was noted to be deficient during a fire drill then a session reviewing the relevant technical and non-technical skills at a simulation centre may be the most effective way of addressing these deficiencies. (An obstetric fire drill is described in Chapter 23.)

Assessment

Assessment of non-technical skills presents many challenges. The most important of these is devising a system that identifies the components of non-technical skills in anaesthesia and is shown to be valid, reliable, sensitive and usable. At the time of writing no such system has been described, although work is ongoing. Experience in other industries has shown that those who will use such a system to carry out assessment, whether for formative or summative purposes , will require training in the use of that system. Calibration of raters, practice at using the system and transparency in the process are necessary if assessment of these skills is to be accepted by their practitioners.

The future

Other industries have shown the importance of training these skills and training those who will be assessing performance in these areas. It is very likely that simulation will have a role to offer in training anaesthetists to use behavioural marker systems to assess the performance of anaesthetists in these skills.

References

1. Helmreich RL, Merritt AC. *Culture at work in aviation and medicine*. Aldershot, England: Ashgate Publishing Limited, 1998.

2. Gaba DM, Fish KJ, Howard SK. *Crisis management in anesthesiology*. New York: Churchill Livingstone, 1994.
3. Jensen RS. *Pilot judgement and crew resource management*. Aldershot: Ashgate Publishing Limited, 1995.
4. Schon DA. *Educating the reflective practitioner*. San Francisco: Josey Bass, 1987.

Further reading

Brannick MT, Salas E, Prince C. *Team performance assessment and measurement*. Mahwah, New Jersey: Lawrence Erlbaum Associates, 1997.

21

Setting up a high fidelity simulator centre

Ronnie Glavin and Nicki Maran

"Well, this certainly buggers our plan to conquer the Universe."

In this chapter we shall look at setting up a simulator centre. At the time of writing there are almost 200 centres with high fidelity simulators throughout the world. Everyone has made mistakes; everyone has done things that they would change. This chapter will not give specific tales of the setting up of centres but will attempt to provide some questions that should be addressed by anyone thinking of setting up such a centre.

The purpose of the centre

What is going to be used for? And who will be using it?

Anaesthetists have played the major role in setting up and running these centres but the educational advantages can apply to all acute clinical specialties. A compromise may have to be reached between what is desirable and what is possible. The idea may be to set up a simulation centre specifically for anaesthetists, but the number of anaesthetists available and the likely sources of funding may modify this. Will the intended groups of learners be confined to trainees, career grades and undergraduate medical students? Will it include other groups of health care professionals such as nursing staff and paramedics? Although it may be possible to approach other groups after the centre is established, one may be able to get a better reward in terms of commitment of both people and resources by approaching key individuals from these groups in the early stages of discussion. It is better to keep one's options open as to the ultimate course of activity of the centre. Involvement of the other groups at the early planning stages, where their representatives can help shape the direction of the centre, is more likely to result in commitment. It can help overcome feelings such as 'Not invented here', that can hinder involvement of the other groups. Similarly, when approaching individuals from other groups, one wishes to get someone who will champion the cause in their own specialty and at the same time contribute to the working group. An approach often has to be made at two levels. It is important to invite contributions from heads of departments, postgraduate deans, directors of training programmes etc. However, these individuals are usually very busy and no matter how well intentioned they are will be able to contribute little to the day to day matters that must be addressed when organising the group. Ideally people who have the respect of their colleagues and enthusiasm for the project are the ones to identify and court.

Theoretically there are 365 days available for teaching. In practice, allowing for public holidays and annual leave, days for course development etc. leave about 200 working days in a centre.

There are several approaches that can be taken. One can design courses de novo and try and sell these to the individuals identified above. Another approach is to look at existing curricula, and show where the simulator can provide teaching opportunities that conventional methods cannot match. This probably makes it more likely for the simulation centre to become attractive to those controlling the training programmes.

Who will use it?

The teaching style of simulation favours small groups of participants on courses taking active roles. It is only by being personally involved that participants can find out what they can and cannot do. This limits the number of participants who can attend any one course. We in the Scottish Centre find that six participants on a one-day, hands-on course is the maximum we can schedule.

One has to look at the possible user groups, the frequency with which they are likely to attend and the money that they will bring with them. Each group that is intended to use the centre will require a set of courses to be developed. As mentioned in the section above, it is probably wiser in the first instance to attempt to fit these courses into existing programmes.

It is unrealistic to expect that the simulation centre will open and be immediately fully occupied. It is much more likely that a few key courses will run and as the profile of the centre is raised other groups will express more interest.

Who will staff it?

Centre co-ordinator/director

In practice most centres will have a local champion, who will play a very active role in setting up the centre and organising the early courses. This individual will require support to run courses. At the outset of a new development the choice of director is crucial. They need to be an enthusiast, and they must have the evangelical zeal and negotiating skills to be able to carry others along with them. Importantly, they must have management skills. There are many strands in the course of a project to bring a simulator centre into operation. These must be co-ordinated and deadlines must be met. It is worth considering having a second in command or even, as we have in Scotland, a co-director. This makes the task of developing a large, expensive project less intimidating and there are definitely sufficient tasks to share between two people.

Faculty roles

The key duties are someone to 'drive' the computer, running the simulation, and someone on the clinical floor helping to 'direct' the scenario. During the debriefing session someone with subject matter expertise should be helping facilitate the discussion. The subject matter expertise will depend on the content of the scenarios. If the main thrust is on the 'technical skills' then someone with specialist experience in that field is needed to give the course credibility. If, on the other hand, the thrust is on non-technical skills then the subject matter expert may be from another discipline. The experience from simulation in

> *To give a course credibility the faculty must include someone with sufficient seniority in the clinical area being taught and this person should usually be from the same clinical discipline as the participants.*

professions other than medicine is that, to ensure credibility, the message must be promoted from the same professional group. For example, a group of clinicians from different medical specialties could function with a clinician and a psychologist but would not be likely to achieve much with a psychologist alone.

Senior faculty usually cannot look after real patients whilst they are teaching. This means that running simulations is expensive for the hospitals that supply faculty. Try to spread your net widely by using faculty from many hospitals.

Similarly, a physician working with a group of nursing staff, or vice versa, is less likely to achieve the desired outcome of the course. The grade of the faculty member can also be important. We would not choose to have a trainee facilitate a group of consultants or attendings. The key to identifying faculty is to look for people who have the time, the enthusiasm and the commitment. In general terms this means looking for people who already have some teaching commitment and attempting to fit that into the simulation centre. This may require that an existing course has to be modified, as discussed above.

The model described is one in which the director has a faculty working below him or her. There is also a need to have someone above the director. It is preferable to have 'a someone' who will act as a supporter of the centre during the inevitable debates about funding, teaching, resources etc. This should be someone who sits on the relevant committees and has so has access to those controlling the supply of these resources.

We also recommend that anyone involved in organising courses and facilitating debriefing is in active clinical practice sufficient to maintain credibility.

Technician

The job description of this post will vary from centre to centre but the basic requirements are:

- Someone to take care of the technical side of the simulator. Simulators have a lot of working components and these require to be maintained, and may require to be repaired or altered during the course of a teaching day. The strength of high fidelity simulators is that participants on courses interact with them. The down side of this is that wear and tear may be higher than with other types of educational apparatus. If the simulator does break down or stop working during a course and cannot be quickly repaired, then the course cannot be completed and the reputation of the centre will not be enhanced.

The technician is a senior and indispensable member of the team — make sure you make a good appointment to this role. Set aside sufficient money to attract someone with the right qualifications. They will not come cheap!

A commitment from a technician employed by another part of the organisation e.g. medical physics may not allow that person to attend immediately for problems. Faculty members who attend the centre infrequently are unlikely to have the experience or expertise required to repair the simulator. In the three and half years our centre has been running we have only had three days where breakdowns in the simulator have significantly interfered with the smooth running of courses.
- Setting up the centre for each scenario. The format that many centres use is one of a scenario in the clinical area followed by debriefing, then next scenario. The centre will have to tidied up, disposables replaced and any necessary props for the next scenario put in place. Some of the more complex scenarios require a level of setting up comparable to an amateur dramatic production.
- Participating in scenarios. We find that we need two members of faculty 'on the floor'. One is the member of faculty who will be directing the scenario on the floor. The other is the person assisting the active participant(s) who will be responsible for getting drugs and equipment. This person obviously has to know where these are kept.

Secretary

The following tasks have to be done:

- Co-ordinating the different courses. Participants have to be identified, contacted and given dates. Faculty have to be informed and recruited etc.
- Paperwork. Course material has to be distributed and evaluation forms collected and analysed.
- Dealing with routine enquiries and correspondence with other centres or with training agencies.

Faculty

The size of the faculty will depend upon the variety of courses and the financial arrangements. Very few centres will have the luxury of being able to 'buy' a lot of faculty time and so will have to rely on goodwill. There are some incentives available for faculty members. In the UK contributions towards teaching at this level may satisfy some Continuing Education and Professional Development requirements and may contribute towards salary enhancement in the form of Discretionary Points. Recruitment, training and development of faculty will be dealt with later.

Funding

The basic outlay is shown in the box.

Capital costs	
Cost of simulator itself	£150,000
Provision of accommodation	£50,000
Refurbishment	£70,000
Additional equipment (including audio-visual)	£50,000
TOTAL	£320,000
Annual costs	
Consultant salary (shared educational directorship)	£66,000
Technician salary	£26,000
Others	£18,000
TOTAL	£110,000

There are several sources that can be approached.

- *Undergraduate Deans*. In the UK this is certainly a route worth trying in the current climate. There is an expansion in the number of medical schools and some increase in the number of medical students in existing schools. Medical school curricula are changing in response to the General Medical Council's document 'Tomorrow's Doctors'. Clinical skill laboratories are also increasing in number. Money may be available and premises may be available as part of these changes. The quid pro quo is that the centre must contribute to undergraduate teaching. As mentioned above, the best approach is to deal with those responsible for the curriculum and point out where simulator courses can fit it. One must temper enthusiasm with a practical, pragmatic approach. Our experience at the Scottish Centre is that simulator courses in the final year of the undergraduate course allow students the opportunity of experiencing clinical cases that are acute, evolving, uncertain and give the students the opportunity of taking responsibility. This can provide a very useful introduction to and preparation for life as a pre-registration house officer or intern. The type of cases that challenge people at this stage can be used and the students can experience the consequences of any therapeutic or interventional strategies.
- *Postgraduate Training bodies*. The questions that have to be addressed are: Do the postgraduate doctors have a training allowance? And if so can this be used to fund costs? Is there access to a budget for capital investment? In the Hollywood film 'Field of Dreams' the voice tells the character played by Kevin Costner "If you build it they will come". This is not the experience of simulation centres. A few enthusiastic participants will volunteer but the majority will only come if there is pressure brought to bear. This is why collaboration with

Consultants will be less willing to attend courses as participants than you expect — be conservative about course fees as a source of income.

those handling training budgets and those responsible for training programmes is so important at an early stage.

> *Do not trust to luck with the funding.*
>
> *It is essential to stay within the proposed budget — so all sources of funding must have given their firm, written commitment before the project starts.*

- *Employing Authorities.* As the concern about patient safety increases, and the liability of hospitals for adverse incidents rises there may be merit in exploring these issues with the chief executives. Concerns about revalidation and preparation for unusual cases may also be relevant here, but unless the support of the consultant body is obtained at an early stage this could prove counterproductive. Many doctors would be suspicious of what appears to be a management led initiative. It makes more sense to enlist support from the medical staff, and then approach management. It is also important to emphasise at this stage that these devices are educational tools to help prepare staff for cases that may be rare, but if not handled properly could result in large amounts of money being paid out. These devices do not have the capability of assessing whether an employee is competent or not.
- *Medical Defence Agencies and Insurance Organisations.* This line has not yet proved fruitful in the United Kingdom but Jeffrey Cooper presented some interesting developments in Massachusetts in this area. All of the attending anaesthetists in the Boston Hospitals Group will be eligible for a reduction in their Medical Insurance Fees if they undertake the Anaesthesia Crisis Resource Management course. As yet, the actuarial evidence behind this move is not in the public domain but it adds to the argument not only for defence agencies but also the employing authorities.
- *Research Bodies.* The actual costs of the simulator and the conversion work required for the site are unlikely to be realised but there may be money available for the employment of a research fellow who may be of help as a member of faculty.
- *Local or Regional Government.* As with the medical defence agencies, there may be government departments that have money that can be diverted towards patient safety. This may be a source for contributions to the capital set up cost.
- *Charitable Organisations.* There are many health related charities which may have a specific interest in such a device. In the UK the British Heart Foundation purchased a Harvey cardiology simulator for every medical school in the UK. Local groups also raise money for local hospital projects and often prefer to put this to some definite capital project.
- *Equipment and drug manufacturers.* Companies are unlikely to meet any major expenditure but may help provide some equipment such as anaesthetic machines, suction apparatus etc. Some drug manufacturers may donate relevant expired drugs. These may help reduce some of the projected costs.

As the number of centres increases, external organisations will become less interested in financing new simulator developments and the money will have to be found entirely from user groups and their employers.

Once the capital costs have been met the annual running costs need to be determined. In addition to running costs (electricity, heat and lighting, disposables etc) there is need for money to replace the simulator. Our annual costs at the Scottish Simulation Centre amount to around £125,000 leaving aside money to replace the manikin. Most of this (£105,000) goes on salaries for the administrative assistant, the whole time equivalent educational directorship and the technician. Courses involving medical staff will usually attract course fees, because either they can afford course fees or have a study leave allowance. This is not the case with nursing staff or paramedics, who will have very little in the way of supporting finances, and are unlikely to be willing or able to commit personal monies to substantial course fees. Estimates of the money available from consultants attending courses should be conservative. This is because they are often reluctant to attend courses, largely because senior staff have a lot more to lose in the way of self-esteem. One has an image of not only what one can do, but also what one is expected to be able to do, and to be confronted with inadequacies in these areas can undermine one's image of oneself. These factors must be taken into account, but unless there is a strong incentive one is not going to be overwhelmed by career grades beating a path to the centre's door.

Where to site it?

The choice of location will, to an extent, be determined by the purpose. This will determine the size of the premises and will also determine how accessible the premises need to be. It is salutary to recall that Sim 1, the first 'modern' anaesthetic simulator, had to be housed in the Engineering Department of the University Campus, because of the amount of space required by the computing function. This meant that it was not immediately accessible to the anaesthetic residents, for whom it was intended and consequently only 13 trainees ever experienced scenarios on it.

Let us consider what is required in terms of the layout.

- *Clinical area*. How many participants will there be? What sort of clinical scenarios will be carried out there? This will determine how much extra equipment may be needed. A larger clinical area may be transformed into a smaller one by using curtains to shield off some parts. Will course participants be expected to view activities from the clinical area? In general large groups of spectators tend to reduce the willingness to suspend disbelief, and scenarios where spectators can contribute are not so popular as those courses where people have hands on experience.
- *Control room*. Most centres have a control room with one-way mirrored glass that allows the occupants to view the scenario while 'driving' the computer. The control

room will require space for the simulator computer and the mixing controls for the audio-visual devices. A telephone connecting the clinical area to the control room is very useful as it allows participants to obtain help, further information and discuss the case with another clinician. It can be used to contact the participant to give further information or add to the general workload. Most visitors attending our centre want to see scenarios in action and the control room is the best location for that, as they can ask questions or raise issues during the running of the scenario. This is very difficult in a cramped control room so our advice would be to allow space for more than just 'driving' the scenario.

- *Debriefing facilities.* These are very important. The key questions to ask are: How close to the clinical area does it have to be? How big does it have to be? The size will depend on the size of the courses and the presence of other adjacent facilities. For example, is there a larger room close by that would serve as a tutorial room for larger groups? Remember that, not only will you have the participants, but you may also have several faculty. We would recommend that you have room for at least 12 people. This will allow two faculty and 10 participants, which is the maximum we would wish to have for hands on courses. Numbers greater than this preclude an active role in the clinical area for all of the participants. Adjacent rooms have the advantage that co-operation between the clinical area and the debriefing room is easier. This is important if both are being used simultaneously. We find that when dealing with large numbers of participants attending on the same day we have to operate in this way to cope with the numbers.
- *Other space.* Key areas are storage, storage and more storage. A lot of equipment may be required such as operating tables, beds, trolleys, dental chairs, resuscitaires etc. A workspace to prepare 'blood' and other 'bodily fluids', keep props and drugs is also to be heavily recommended. Video cassette tapes, completed evaluation forms, copies of course material etc. will require to kept plus databases of participants etc.
- *Changing rooms, toilets etc.* Is the centre going to be self-contained or will it be housed in a location where these things are available? It will also depend on the nature of the courses. Suspension of disbelief is more likely in a theatre environment if people are dressed appropriately but this may not be so necessary for those courses where the clinical area is a ward space or the ER etc.
- *Office.* Very few centres will have the luxury of being able to design a centre from scratch. Most centres have to be accommodated into an existing structure. It is obvious that a secretary will need somewhere to carry out her duties but we have also found it very useful to have a location where some privacy is possible. This has been helpful not just for sensitive telephone conversations but also to speak to participants.
- *Potential for expansion.* Although the first priority is to establish a centre, and therefore short-term considerations will take priority, it is important to not to ignore longer-term plans. Most centres will be shoe-horned into existing premises but

there may be some choice as to the locations on offer and one of the considerations should be where could we expand if we need to? Increases in demand may need anything from an increase in storage space to an increase in the clinical simulation area. At the setting up stage all one needs to consider is what options are possible? Could staff in adjacent premises move if necessary? Or are there specific items of equipment or geographical considerations that would not allow such relocation. Even if they can move would there be somewhere for them to go? Naturally, a sense of tact and diplomacy are necessary when handling such issues.

How long will it take to set up a centre?

There are several phases.

- Securing the financial and political support.
  This is usually the longest phase and by the time that the important committee meetings have taken place several years may have elapsed.
- Building work and conversion of facilities.
  This phase is usually down to weeks. Inevitably it will take longer than the estimates given (a situation familiar to anyone who has had to use builders domestically).
- Training and preparation of course.
  This phase can begin during the above phase and will to an extent depend on the previous experience of faculty. It is worth allowing two months for this phase, which should give time to develop and rehearse course and give time for participants to come (study leave arrangements etc.).

Get a professional project planner to advise you about the organisation of your project plan.

22

Practicalities of simulation: adding role-play to low and intermediate fidelity simulation

Aidan Byrne and Matthew Checketts

> *Simulation requires the participants to suspend their belief in reality.*

> *You must select a type of simulation that is appropriate to the learner and the lesson.*

This chapter covers the practicalities of running simulations. The introduction covers the reasons why certain aspects of each simulation are important. The later sections detail certain types of simulation and how each is constructed.

A crucial difference between simulation and other forms of education is that simulation requires the participants to suspend disbelief. So, for example, while an essay may require a trainee to list the actions they would take when confronted by a breathless patient, in a simulation they are asked to pretend that such a patient exists and to physically perform the actions. While most of us pretend to be rational beings, the power of television and radio to provoke interest and emotion are constant reminders that humans have a huge innate capacity for drama. Once disbelief has been suspended, trainees can then learn, not only isolated facts or solutions, but how situations 'feel' and the way their actions can be co-ordinated. After all, most medical problems are not solved by a single act, but require the practitioner to diagnose, initiate treatment, inform the patient, involve and co-ordinate other workers, as well as reviewing the effects of their own actions. Although a minority of trainees will be unwilling to suspend disbelief, most will require little encouragement to fully engage in the process.

The educational aims of simulation

Simulation is an educational tool. The point is to teach a lesson that has been planned.

Simulation is an involving and often entertaining process. While the realism and excitement it generates underpins its effectiveness as an educational tool, it also means that there is a danger that simulation can become an end in itself. It must always be remembered that the aim is to improve the skills of the participants. As is detailed in other chapters, each simulation should take into account the skill level and experience of the trainees. For example, undergraduates exposed to a full scale, high-fidelity simulation of major trauma are likely to be overwhelmed by the experience. At best they are unlikely to learn effectively, and at worst the experience may result in real psychological trauma. In contrast, experienced practitioners may feel insulted if asked to practice, or even worse, have their clinical competence assessed during a low-fidelity simulation.

The aspect of performance that is being taught is crucial to the form of simulation that is chosen. For example, if the intended educational aim is improving the subject's knowledge of the procedure for a cardiac arrest, then a role-play may be the most appropriate method. In contrast, if the aim is teaching the skill of inserting an epidural needle, then the simulation must include a high fidelity interactive model of the lumbar spine that includes the use of real equipment and one that provides appropriate tactile feedback.

The initial phase of planning can also be used to investigate the range of skills needed to perform the task under scrutiny. For example, a teaching session based on inserting epidural catheters for post-operative analgesia could simply focus on the skill of needle insertion. Depending on the group being taught, other skills that may need to be addressed could include selection of appropriate equipment, aseptic precautions, location of an appropriate vertebral interspace, obtaining informed consent, communication skills and what to do if the patient cannot tolerate the procedure. For example, there is little point in ensuring that a trainee can insert a needle into a patient when they have little or no idea of how to select appropriate patients for such a procedure. Teaching trainees about patient selection and communication helps trainees to remember about the skills they have practiced as they are then in an appropriate context.

While inexperienced trainees are often taught isolated skills with basic equipment, more experienced trainees can combine multiple elements. For example, an obstetric anaesthetist could be exposed to a simulation combining the skill of inserting an epidural into a high-fidelity, lumbar spine simulator, discussing possible complications with a 'patient', played by an actor, plus having to supervise a 'nurse', played by an actor. Ultimately, a high fidelity simulation can introduce such elements as a 'husband' who collapses during the procedure and starts to fit.

Ideally, the simulations should progress with the skills of the trainees, so that, for example, a trainee being taught awake, fibre optic intubation could, after the initial book learning, learn the physical skill of handling a fibreoptic endoscope in a simple training box, then learn how to negotiate the normal airway anatomy with an airway model, then learn how to deal with mucous and blood in a high-fidelity endoscopic trainer. Finally, the same endoscopic trainer could be combined with an anaesthesia simulator so that the trainee not only has to cope with the airway, but also has to recognise and treat such problems as hypoxia or cardiac arrythmias. Such an approach is commonplace in aviation where pilots will start with book reading and lectures and then progress onto part-task trainers, low-fidelity and finally, high-fidelity simulation. The cost of such training schemes is obviously much greater than medical education budgets currently allow.

Directing a simulation is a skilled job requiring preparation and practice.

Running a simulation is a skilled job. The tutor needs to combine a sense of theatre, an understanding of the process being simulated as well as an understanding of how the trainee is reacting and how best they should be handled. The tutor needs to find a compromise between an unrealistic, unthreatening simulation (good for a novice, but boring to an experienced clinician) and a complex simulation involving multiple staff, which may cause inexperienced staff to develop signs of acute anxiety and possibly lasting psychological damage. Any novices are best advised to attend some established courses and to see another group's ideas in practice before starting their own.

Curriculum

Some items in the curriculum are very suitable for teaching by simulation and role-play.

Medical postgraduate exams have traditionally not had an explicit curriculum, with the result that trainees have often been surprised at examinations by questions that they did not expect to have to answer. In contrast, in aviation, the form and content of certification examinations are entirely explicit and the correct response to any problem is not in doubt. More recently, medical school curricula have become much more explicit.

Ideally, simulation should be incorporated into the curriculum so that all the participants know exactly what they are expected to know and what they are expected to do, as well as how they will be assessed and what constitutes an adequate performance. At present, simulation has not advanced to the point where it can be used in the formal assessment of competence for medical practitioners. However, the stress of simulation can be greatly reduced by ensuring that each participant understands the process and knows that they will not be tricked or humiliated. This process is much more difficult with simulations performed in isolated centres as participants are likely to have travelled from a number of centres.

Scripts

In their simplest form scripts outline the problem, the solution and the consequences of each possible decision. For example, a patient has an episode of pulseless ventricular tachycardia. If the trainee defibrillates correctly, the patient gets better, otherwise the patient dies. While this approach may be suitable for the most basic simulations, more complex simulations require a more sophisticated approach. This may be because more than one solution is appropriate or because there are a variety of partial solutions, which the trainee might use in the initial stages of the problem. There may also be some strategies which are wrong, but which it might be useful for the trainee to experience. For example, in a simulation of a blocked endotracheal tube the correct solution is to replace the blocked tube, an alternative might be to remove the tube and then resort to bag and mask ventilation. However, a trainee who removes the blocked tube, checks the patency of the airway and then replaces the same blocked tube in the airway, may learn a valuable lesson.

Scripts therefore need to incorporate the facility for the trainee taking the correct action as well as a variety of other responses. However, trainees are human beings and are also likely to respond inappropriately and sometimes in frankly bizarre ways. Scripts may also need to be designed to cope with some of these problems and experience in simulation design is extremely valuable here. Failure to anticipate the actions of trainees may result in the simulation having to be stopped, for explanations to take place, with a resultant loss of realism and continuity.

In role play, the script is likely to be similar to the script of a play, a series of 'lines' for the tutor, delivered in response to the actions of a trainee. In medium fidelity simulators the script may be encoded into the computer in the form of a 'node and branch' series of links, again linked to the actions of the trainee. Although many high-fidelity anaesthesia simulators have complex pharmacological models that respond to the actions of the trainee, scripting is still necessary in practice. Those using these simulators are usually forced to adjust the running of the software on a regular basis during each simulation. This may be either to allow for factors which are not included in the model or correcting an unexpected event that the software has simulated.

The script can also be used by the faculty to list all the equipment that needs to be provided, to set the scene, to position each participant at the start and to add important details of each individual's skills and motivation.

In addition to the master script for the simulation director, it is useful for each participant to have an individual script. This would include such details as their role, past experience, the clinical details of the patient, details of the clinical environment and available services, as well as a description of the role and training of the other staff present as appropriate. For example, the script may allocate one member of staff to act as a theatre porter who is able to help with physical actions such as moving the patient, but who is completely ignorant of medical terms. So that when asked to fetch a defibrillator, they bring back an ECG machine.

The use of well designed scripts means that trainees can understand their role and ask appropriate questions before the simulation starts. This is invaluable as it avoids trainees becoming frustrated and angry at the end of a simulation because they did not realise that resources that they needed were available all the time. As in almost all aspects of life, detailed planning and practice is vital to overall success.

Running simulations

Ice-breaking

Most learners are very frightened at the prospect of their first attempt at simulation.

Simulation must not result in the learner being humiliated.

Simulations can be formidably intimidating to the uninitiated, especially in a remote centre when the participants may be unfamiliar to each other. The result can be an unwillingness to take part in simulations and a deathly silence during debriefing sessions.

This reluctance can be reduced by good pre-course education, principally written information describing the form and aims of the course, as well as starting the course with a 'warm-up' session where trainees can take part in a simulation in a non-threatening way. For example, this can be a simulation where all the trainees are given a role with an extreme personality, such as a maniacal surgeon or a depressed nurse, in a short session where the clinical outcome is irrelevant. An alternative is to run a 'quick-fire' simulation, where each participant only gets 30 seconds to solve the problem before handing on to the next trainee. The aim should be for all the trainees to take an active part in the simulation and for lots of errors to be made so that everyone can be assured that their self-esteem is safe. While laughter is an essential component in getting trainees to relax and participate, care is needed to avoid trainees becoming too dramatic, with subsequent simulations degenerating into amateur dramatic productions.

Courses should also be designed so that initial simulations have obvious, simple solutions with subsequent simulations becoming progressively more complex and challenging. An example would be to start out with a simple bradycardia that responds to a dose of atropine, moving through episodes of bronchospasm, ending up with a severe episode of anaphylaxis requiring the intervention of several staff.

Initial briefing

If multiple groups are present, they should be separated so they can be briefed separately. If written sheets are available, the trainees can be asked to read through them and to ask questions about their role, experience and the role of the other participants.

Preliminary briefing is important so that everyone understands what is happening and what their role will be.

A couple of minutes of reflection is useful to allow all those taking part to get into role. Once people are positioned, it is important to give a very clear signal that the simulation has begun. If this is not done, one or more of the participants may fail to realise that the simulation has started with resultant disruption.

Running a session

All simulations require a 'Director', clearly identified as the person in charge, who can rapidly assist trainees who have minor problems and who can ensure that the simulation runs well. While the director can increase the stress levels of participants by talking anxiously about the poor state of the patient, or the lack of blood pressure, this often leads to trainees becoming overanxious. Where inexperienced trainees are taking part, the director may need to make frequent interventions to reassure the trainees that what they are doing is correct and perhaps even to correct any misconceptions that have arisen. Even very experienced trainees may require some reassurance if they are to fully engage in the simulation.

Stop the simulation prematurely if it seems that there is unlikely to be any further learning.

Where the trainee is performing poorly, the director will need to decide at which point the simulation should end. Once a trainee has failed to identify the problem, or has said that they do not know what to do, it is important to stop the simulation as there is unlikely to be any further educational gain. Depending on the aims of the course, it may be appropriate for the director to end the simulation successfully, even where the trainee has performed poorly. This is a contentious area, for example, in civilian aviation simulators, simulations never end with the loss of the aircraft and the emphasis is on increasing the confidence of the crews. In contrast, in military aviation simulation, aircraft damage or even total loss are not unknown and criticism of the staff can be harsh.

Ending a session

Once the session has ended, getting all the participants together to discuss their performance is vital. Watching the videotapes of the simulation can help discussion. A problem with many simulations is the vast quantity of educational opportunities that even short simulations can generate. This can provide wonderful material for discussion, but if simulation is to be used effectively it is important for the director of the simulation to ensure that the trainees focus on appropriate aspects. For example, experienced practitioners often focus on technical aspects of their performance, such as the type of drugs used or whether a particular action was 'right' or 'wrong', instead of communication and team working. Debriefing is a specialist skill that can add hugely to the value of simulation, but it is a skill which tutors need to learn and to practice.

The components of simulation

The setting

Simulations can be run in dedicated simulation centres in remote locations, in training areas within the hospital grounds, in training areas alongside clinical areas, or in clinical areas themselves. Remote areas have the advantages that the trainees are in an educational setting and cannot be bleeped or called back to work. As the trainees are likely to have travelled to the centre and expect to learn, they are likely to be prepared to participate and learn. In addition, all aspects of the environment can be controlled, so that staff can arrange scenarios involving events such as fire alarms or equipment failures that would risk disaster if staged in a clinical environment. As well as the trainees being relaxed, the trainers are also helped by the lack of distraction and by working in an educational environment. The downside of dedicated educational centres is the cost in terms of staff removed from clinical duties, the cost of the buildings and equipment and travelling, food etc. There is also the possibility that removing the training duties from the clinical workers may result in training being seen as something not relevant to the real world, which may be reinforced by the clinicians left at the workplace if they are not adequately involved in the training process.

In contrast, working in clinical areas has the advantage of facilitating group working, so that entire teams of nurses, doctors, technicians and support staff can be trained together, which may result in improved learning as well as fostering team spirit. In addition, any local problems, such as deficiencies in equipment, personality conflicts or poor lines of command or communication will be identified and can be corrected. However, arranging for the cancellation of elective work for training may be unpopular. If staff are absent for training it is rarely noticed, but if staff are present in, for example, a theatre suite and patients are waiting for surgery, staff will invariably be pressurised into abandoning the training. Taking non-functional training devices into clinical areas may also be dangerous as they may find their way into general use with disastrous consequences.

Types of simulation

Tutor guided role-play

There are no physical representations present and the simulation progresses as an exchange between trainee and tutor, perhaps with others contributing. The advantage of role-play is that it is cheap and instantly accessible. For example, if a trainee experiences a difficult clinical case, the tutor can wait until there is relative calm and then talk the trainee through the problem.

Obvious problems with this form of simulation are the lack of physical realism and the inability to practice physical skills. There is also a danger that the session can become a jousting match between the trainer and trainee. If this occurs and the trainee fails to find the correct solution, they may feel that they have been misled by the trainer and that they have been deliberately put down. However the proximity of teaching to real world experience can provide some valuable teaching as, for example, a clinical difficulty can be followed by a session starting "Well done, you managed that problem well, but what would you have done if your first line treatment had failed?" Any learning in this context is likely to be well received and linked directly to clinical practice.

An example of role play:

A pregnant patient is about to undergo an emergency caesarean section. The trainee describes a rapid sequence induction and gets as far as laryngoscopy ...

Trainer — Sats are now 95%, you look into the pharynx and can only see the epiglottis.
Trainee — I would reposition the head and use a longer blade.
Trainer — You look again and the view has not improved.
Trainee — I pass a gum elastic bougie blindly and railroad a tube over it, connect it to the circuit and check for carbon dioxide.
Trainer — Sats are now 85% and the trace on the capnograph remains flat.
Trainee — I would ask my assistant to remove cricoid pressure to get a better view.
Trainer — As you look again the pharynx fills with stomach contents, Sats are now 70% ... etc.

In addition, verbal exchanges can provide an essential part of more complex simulations. For example, in 'Moulages', the 'patient' is usually represented by a manikin, with the clinical signs and other data provided by the trainer. Even in high-fidelity simulations, clinical signs such as cyanosis can only be confirmed verbally by the trainer as current technology does not allow such signs to show on manikins.

Simulated patients

Trained actors can stand in for real patients. They can provide conversation and real clinical signs. However, participants need to have both theatrical skills and medical knowledge so that they provide realism without misleading the trainee. Many medical schools now use professional actors who have been trained to provide standardised histories. These 'patients' provide an extremely valuable resource, as their skill means they can be used in assessments with a high degree of validity. While clinical signs such as cyanosis and pulselessness are difficult to fake, other signs such as abdominal guarding and rebound tenderness can be simulated effectively. In some centres, part-task trainers, such as artificial skin, are attached to actors so that a trainee can, for example, suture a 'laceration' on the arm of the 'patient'. This ensures that the trainee not only learns the skill of suturing, but can also practice the skills of communication, aseptic technique, safe disposal of sharps, use of local anaesthesia, all in a single session.

Simple manikins

Manikins can take the place of the patient and are commonly used for resuscitation training. The basic resuscitation manikin with the usual features of a palpable pulse, a facility for defibrillation and ventilation can often be modified to enhance simulations. For example, intravenous access arms can be connected to reservoirs to allow the infusion of fluids and injection of drugs, a speaker can be placed inside the head of the manikin to allow the manikin to speak and additional padding can be added to simulate pregnancy, obesity or oedema. Some manikins have arm inserts that contain veins for cannulation or venepuncture. Clothing can be added to aid realism and to hide simulated injuries. A simple approach is to fix labels to the manikin so that, for example, when the arm is exposed the trainee can read, "arm is in an abnormal posture with a small wound from which the end of a bone can be seen protruding." Theatrical suppliers can often suggest products such as fake wounds and blood that can be very effective.

In the last few years the varieties of manikins in commercial production has increased rapidly in parallel with the different features available. Many of the more sophisticated models come complete with computerised monitoring and the ability to generate script based simulations. While many of these models can provide a suitable basis for simulation based training, the use of computers and electronic features may mean that it is difficult to use them in roles other than those for which they were designed.

Monitoring

Real monitors are expensive to purchase and require modification if they are to be used as part of a simulation. High-fidelity simulators use this approach, which is highly effective, provided the trainees are familiar with the type of monitors used. The ACCESS and SimMan simulators use a computer to generate a cartoon-like picture of anaesthesia monitoring, which is cheaper. A very low cost option is to print an image of the monitor display onto cards, which can be placed in front of the monitor screen.

As the condition of the patient changes, the card can be changed to show different values appropriate to the condition of the patient. The lowest tech option is to have the tutor call out results at the request of the trainee.

Unfortunately, monitoring errors, such as fixation on one monitor or failure to use the information correctly are a rich source of mishaps. Providing the information verbally, or by moving printed cards, at regular intervals, acts as a prompt for the trainee to re-evaluate the data and will prompt them to check the data if they failed to understand it clearly. The presentation of data in such an unequivocal way may greatly reduce fixation error and falsely improve the performance of trainees.

Investigations

Investigations such as arterial blood gas estimation can easily be provided verbally by the tutor. A more realistic approach is for the tutor to write the result on to paper and pass it to the trainee, or to ask the trainee to telephone a number, where an accomplice is waiting with the result.

It is important to ensure that the results do not appear out of thin air as soon as they are requested. Time should be allowed to take the sample, for transfer to the analyser, for analysis to take place and for the result to be transferred back to the trainee. It can be extremely educational for samples to be lost occasionally or for technicians to refuse to process the sample because the form has been filled in incorrectly.

In the same way, stocks of standard X-rays, ECGs and other results can be kept in stock and provided to the trainee after an appropriate time has elapsed.

Other staff

Communication with other staff is a skill that is rarely taught in postgraduate medicine. Simulations run in normal clinical areas have the advantage of involving the full range of staff, so that lines of communication can be tested and problems identified. This can be particularly valuable in that any problems identified are relevant to everyone present and solutions can be identified and put into place immediately. Personality clashes can be particularly disruptive in clinical environments and can be particularly difficult to identify and solve. Especially where simulations have been videotaped, it can be very instructive for the staff to describe exactly what they thought they said and then to listen to what they did say.

Staff from outside the immediate clinical environment can also be involved in the simulations. For example, in a simulation of major haemorrhage, dummy samples can be sent to the blood bank. The blood bank staff can then wait for the appropriate time and then send back bags full of red liquid. This can identify problems in portering services, communication and in organisation of the hospital. These might include the siting of the blood bank relative to the clinical areas or the quality of the portering staff.

Alternatively, other tutors or actors can play the roles of other clinical staff. This has the advantage that those chosen can have particular characteristics, such as arrogance or inefficiency that can be identified and discussed in the debriefing. They can also help to steer the simulation in the desired direction if the trainee appears to be failing to address the intended problems. The main problem is that such staff need to be familiar with both the simulation techniques being used and the particular scenario being run. It is often difficult to keep such staff available.

Equipment

Simulation provides the ideal environment to try out new pieces of equipment that are only used on an infrequent or emergency basis. Often the educational aims of these simulations are limited to the physical skill of handling the equipment and the simulations are correspondingly limited. For example, the skill of manoeuvring a fibre-optic intubating bronchoscope can be practised by passing the endoscope through a cardboard box containing a number of baffles, boxes or other detritus which have been fixed at strategic points. A trainer can easily provide a background of coughs, descriptions of ECG abnormalities and variations in oxygen saturation during the passage of the endoscope.

Other examples include deactivated defibrillators which allow realistic defibrillation training to take place without the risk of electrocuting any of the trainees or trainers or using a variety of foodstuffs to mimic difficult to mix pharmaceuticals such as dantrolene.

Summary

Simulation can form the basis of an exciting and educationally effective teaching programme. However, the prime consideration must always be the educational aims of each session. Once these are established, then the appropriate range of actors, equipment, setting etc can be identified. In general, the level of simulation required is determined by whether or not the participants are able to suspend disbelief or not. In most cases, effective learning can be achieved with fairly simple levels of equipment and staffing.

23

How to organise a major obstetric haemorrhage 'fire-drill'

Rachel Walpole and Vicki Clark

Simulation and role-play can be used in a number of situations so that the whole hospital can practice together. As an example we will describe the procedures for dealing with major obstetric haemorrhage. A scenario that is commonly practiced is the whole hospital major disaster plan, and a similar approach can be used for other emergencies that require the co-ordination of teams, and where prompt correct treatment makes a difference to outcome. Examples are major burns, simultaneous reception of two or more patients with major injuries, inhalation injury from chemicals and biological or radio-active contamination.

Life-threatening haemorrhage in pregnancy occurs in 1:1000 deliveries in the UK;[1] this means that a large obstetric unit can expect to see only a few cases every year, and an individual member of staff will be involved in such a case very infrequently. Many members of staff working on Labour Ward, in particular anaesthetic and obstetric trainees and junior midwives, are not there on a permanent basis, as they are rotating through different areas of the hospital. This can result in a lack of familiarity with protocols, particularly those that are infrequently used. In turn, this can cause delay in

In a 'fire-drill' the whole sector of the hospital that is involved in the patient's care plays out the management of a simulated emergency in the setting of the real workplace.

an emergency. The Confidential Enquiries into Maternal Deaths (CEMD) suggested that one way to improve this situation is to practise for major haemorrhage with a mock 'fire-drill'.[1] For the last two years we have organised regular mock drills for major obstetric haemorrhage in our hospital.

In a 'fire-drill' the whole sector of the hospital that is involved in the patient's care plays out the management of a simulated emergency in the setting of the real workplace.

Preparation

Suitable scenarios:

Placenta praevia. A 24-year-old primigravida with a known low-lying placenta presents at 36 weeks gestation with painless vaginal bleeding.

Placental abruption. A 29-year-old para 1 + 1 presents at 32 weeks gestation complaining of sudden onset of abdominal pain, shocked, with no vaginal bleeding and a very poor cardiotocograph trace.

The bulk of the work involved in a drill is the preparation. However, it is important to keep the plans a secret from all but a few key people, or you will find that the duty team have been doing some preparation of their own! Identify a day well in advance and, if possible, arrange for it to be free of elective work so that there is a relative abundance of staff. Then, decide on the scenario you want to simulate.

The initial scenario can then be developed on the day, depending on the speed and quality of resuscitation, but is likely to include coagulopathy and/or uterine atony necessitating hysterectomy.

Once the scenario is decided, the preparation then falls into three categories: personnel, equipment and communication.

Personnel

First, recruit an assistant. There is an enormous amount to be done, and you will need some help. The ideal assistant (assuming the organiser is an anaesthetist) is probably an obstetrician, either consultant or senior trainee. A cross-specialty organising team will promote the view that this is a Labour Ward, rather than an anaesthetic drill, and will provide useful insight into the obstetric management and thought-processes. During the drill itself, your assistant can stand back and observe what is going well or badly and make notes for the debriefing. You will be too busy developing the scenario and communicating with the team to do this.

Next, recruit a patient. The best patient will have some medical knowledge in order to describe symptoms and display signs, but will be unknown to the Labour Ward team. A medical student is usually a good choice, but an anaesthetic SHO is an alternative.

Ask your patient to pad her abdomen appropriately, and to wear old clothes, as they are likely to get 'blood'-stained! Brief the patient well as to her story, and encourage her to throw herself into the part. It is very unusual for any of our real patients to come to the Labour Ward alone, so recruit a father for the baby, a male SHO or ODP are ideal.

Clearly there is no need to select a team for the drill, as it will be the duty team for the day, and the surprise factor is of the essence. Personnel on duty should be brought in to help with the emergency as they would be in the genuine situation.

Equipment

Produce a sheet of paper detailing instructions for participants so that everyone is clear about the rules of the drill. This should be handed to each team member as they enter the scenario. A suitable format is shown in the box.

> This is a mock drill.
>
> The 'patient' is not really unwell, so any invasive procedures should be simulated. Say aloud what you are doing, e.g. "I am taking blood for a full blood count." Everything else that you might do in this situation should be carried out exactly as normal, i.e. once blood sampling has been simulated the tubes and forms should be filled in and sent to the laboratory. Drugs and intravenous fluids should be given into a bucket beside the patient.
>
> Dr X is coordinating the drill, and will give you more information as the scenario progresses. If anything is unclear at any time please ask.

Each participant should be given a badge so that there is no ambiguity about the role they are playing, particularly if a regular member of staff plays the 'partner'.

Make a set of notes for your patient. She will need an obstetric history and a scan result if appropriate. A set of sticky labels with the patients name, number etc. makes the management of any emergency easier!

Prepare blood and products. With a pot of food colouring and a syringe and needle you can rapidly transform 500 ml bags of crystalloid fluid into blood products: red for red cell concentrate, and yellow for plasma, platelets and cryoprecipitate. You will need 17 red bags (2 O-Negative and 15 group-specific units) and 6 yellow bags.

Fill some blood tubes with the mock blood.

Prepare a 'bloody' sheet for your patient to arrive on if appropriate and some 'bloody' swabs for theatre.

Communication

Ideally the drill should come as a surprise to all members of the Labour Ward team. However, in the real world, it is polite to give advance warning to the Consultant Anaesthetist and Consultant Obstetrician on duty that day, emphasising the need for

A 'fire-drill' rehearses and tests communication within the department and team working of the whole team — not just the medical team. It is inclusive of everyone including porters, care assistants and telephonists.

secrecy from other team members. An alternative is to warn everyone by memo that the drill will occur at some point in the next month, and then inform no one of the exact date.[2]

Smooth running of the drill requires the cooperation of a number of non-Labour Ward staff, and it is vital that they are not subject to the surprise element:

- *Blood bank.* Ideally you would like the Blood bank to receive the sample, wait an appropriate time as if matching the blood, then despatch your red and yellow bags with Blood bank labels stuck on, and a Blood Issue form to accompany them. This allows proper blood-checking procedures to be followed at the bedside. Some departments may be more willing than others to cooperate with this, but good communication with a Consultant Haematologist, well in advance of the day, will maximise the chances of help. If encountering difficulties, it may be worth reminding the Blood bank of the importance that the Confidential Enquiry into Maternal Deaths places on their participation in these drills.[1]
- *Haematology and Biochemistry laboratories.* These laboratories will be receiving samples of coloured water during the drill, so it is essential to let them know in advance what is going on. They may agree to ring back a pre-arranged set of blood results at an appropriate time interval after the sample is received. Again, early communication is the key.
- *Portering staff.* The porters play a key role in any major haemorrhage situation, and it is important that they take their part in the drill in the same way as other members of the team. However, it is diplomatic to warn the head porter that his staff will be taking part in a training exercise.

The day itself

The 'fire-drill' should be unexpected — but wait for a quiet time before starting!

On the day of the drill, put the two units of O-Negative 'blood' into the Labour Ward fridge, and deliver the rest of the mock products to the Blood bank. Make sure buckets are available for your patient's intravenous fluids to run into. If you have them, place the teaching kit for arterial and central venous pressure lines in the cupboard with the 'real thing'; this cuts down on the cost of the drill. Any dummy equipment must be clearly marked as such to avoid it being used for real patients. This particularly applies to O-negative 'blood'.

Begin the drill when Labour Ward is reasonably quiet. As each member of the team is called into the scenario, explain that it is a drill and give them a copy of the instructions.

At intervals, usually five minutes, but more frequently if asked, inform the team of the patient's current vital signs, blood loss etc. If the patient is in theatre, a sheet of paper showing latest vital signs can be stuck on the monitor display, and blood loss written on the board. When a blood sample is sent away, an appropriate time interval should elapse before the results are communicated. You can suction 'blood' from the buckets to make the blood loss look realistic.

Inevitably you have to make it up as you go along from this point, guided by how the team is handling the resuscitation.

At the end of the drill, send the team away for a cup of coffee and to calm down, but ask them to return in half an hour for the debriefing. At the debriefing, remember the sandwich approach: praise, criticise, praise. Ask for feedback: how did they feel it went? How could we improve our response to a real emergency? How could we improve the simulation?

After the debriefing, remember to thank and give feedback to all the supporting team members, particularly the blood bank. Retrieve any unused mock blood (do not forget the O-Negative in the fridge!) and store it away for next time.

Running major simulations has dangers. One of the editors of this book has twice been involved with major disaster practices where members of the team sustained real injuries because passers-by thought the emergency was real and panicked. On another occasion a simulated 'railside' casualty was given a real dose of pethidine by a passing general practitioner and had to be admitted for observation.

> *As with all simulations, debriefing and discussion is an important part of the learning. All those groups of workers who have been involved should get feedback.*
>
> *Complete the cycle by making changes where systems failures have been unearthed — that is a major purpose of the exercise.*

Obstetric 'fire-drills' in our own unit and others have revealed a number of problems, in particular lack of familiarity with protocols, equipment problems, and inadequate communication between team members.[2] Hopefully with practice and by acting on and communicating widely the lessons learnt from each drill, we can improve our handling of a real emergency. In our own unit the feedback from participants in mock-drills has been overwhelmingly positive and we are now planning to expand our drill 'repertoire' to cover eclampsia and cardiac arrest.

References

1. *Why mothers die.* Report on Confidential Enquiries into Maternal Deaths in the United Kingdom. Department of Health, 1998: Chapter 4.
2. Gaunt A, Kirby S, Holdcroft A. Simulation of obstetric anaesthetic emergencies: an analysis of ward based scenarios. *International Journal of Obstetric Anesthesiology.* 2001;10:246.

Appendix 1

A guide for teaching moderators how to conduct a problem based learning session

1. Explain what PBL is.
2. Describe the components of your PBL program.
   - Active learning
   - Small group
   - Learner centred
   - Case based
   - Problem orientated.
3. Discuss the philosophy of PBL.
   - 'Learner centered' not 'teacher centered'. PBL allows the learners not the teacher to control the agenda.
   - Active learning where time constraints are more relaxed. This is the dilemma that is confronted in PBL.
   - Traditional PBL requires student investigation and reading. Will your students take the time? Providing the learners with reprints or text assignments is an instructional strategy to save time.
4. Discussion of the objectives of PBL session.
   - Produce learning in discussants/learners.
     - Content
     - Problem-solving
     - Judgment
     - Differential diagnosis.
   - Inquisitive, thorough, and rational discussion of a case and the problems it presents.

- Penetrating risk/benefit analysis of clinical options.
- Opportunity for all participants to contribute to the discussion.
- Aura of satisfaction and contentment at conclusion.
- Group should feel 'good' about the discussion.
- Aim for a high 'happiness index' at the conclusion.
- Participants who have worked hard in preparation for the session should be allowed to feel that they have an opportunity to present their information and strategies.

5. Discuss the role of facilitator.
   - Your creativity is the boundary of your limits, in a spontaneous environment.
   - Each group will be different and present new challenges and opportunities.
   - Not information provider: In a lecture the teacher is an information provider, but in this format the faculty is moderator/facilitator. This is why the case writer expended so great an effort in creating a case that will take the learners to learning issues that the case writer has set for them!
   - Moderating and facilitator skills are used to produce learning in the students.
   - Allow the students to experience the joy of insight and understanding in the course of their active learning.
6. Moderating/facilitating skills.
   - Moderating/facilitating skills are different from traditional teaching skills. This is learner-centered education.
   - Is your PBL session solitary or a series of PBL sessions. In the ongoing format you want to know as much as possible about the learners so that you can be more effective to them as individuals. In a single session the challenge is to forge a group of strangers into an effective discussion group.
   - 'Opening set': creating the right learning environment for a small group session among strangers.
     - 'Introduction: breaking the ice'. Have the participants introduce themselves and tell a little about themselves.
     - Do the learners know each other? What do they bring to the session? Questions, theories or thoughts generated from the case.
     - Establishing the environment. Smile and encourage the thought that you are confident and will provide the group with a wonderful learning experience. We are going to have fun! This is the attitude you wish to convey! Remember your body language and facial expression is a powerful mood setting device even before you utter your first words! Establish the ground rules. Use first names. State that you will try not to be an agenda setter or information giver. Pay attention to non-verbal cues from the participants and be aware of your own non-verbal behavior.
     - Clues form the opening set period. Have they read the case? What are their feelings about the case? Are they excited about the session?

7. Promotion of discussion. Dilemma of having freedom of student control of learning agenda versus covering content of the case as moderator/facilitator.
   - Roles and responsibilities within a group.
   - Timekeeping.
   - Decorum. Encouragement of all to talk, holding down speechmakers, modulating overbearing group members.
     - Information providers
     - Agenda setters
     - Group consensus builders
     - Silent majority
     - Establishing collegial, non-threatening environment.
   - OK to take 'time out' from the case to talk about process, i.e. is the discussion covering the molecular mechanisms of the drugs we propose giving to the patient in the case? How can we be more effective in this group task?
   - Times for intervention.
     - Group is far afield.
     - Group is off base, e.g. too much time on psychosocial aspect of the case rather than content issues! Need for greater analysis in discussion, getting the explanation to the molecular level or debate on data in the literature! Moderator intervention to promote focusing of the discussion.
     - Controversy too heated (seldom occurs), becoming personal — need to defuse it!
   - Do not miss cues to change agenda: much better than introducing a new topic de novo, but opportunity to change the topic in context of earlier and future discussion.
   - Cover the content.
   - Do not permit group to become bogged down in peripheral issues!
   - Encourage group to use the flip chart. Written form forces organization and systemization of thoughts with clearer explanations!
   - Session is boring. Liven things up by creating controversy, move to a role-play, convert to oral board examination format, create activity!
   - Monitor the pace and tenor of the discussion and make decisions about intervention consciously!
   - Monitor individual members of the group: is this one afraid to speak out? Are most of them bored? Look at their eyes. Are they bright, attentive and excited or dull and bored?
   - Redirecting a floundering discussion. Some times the discussion will seem to be floundering: You need to recognize this and inject some activity and provide some direction with direct and leading questions. Some groups may not have prepared satisfactorily. Do not despair! Provide them with direction and even some facts. Make the best of the situation. You can convey disappointment,

but try to salvage the discussion by at least demonstrating why the case and problems within the case excited you.

- Questions. Questions are your tools to sculpt, mold, enliven and create the nature of the discussion.
  - Open-ended.
  - Focusing.
  - Promote higher cognitive levels.
  - Ask if the discussion is on a factual level or should the group be analyzing, synthesizing and evaluating the elements of the problem?
  - Ask for facts.
  - Ask for philosophy.
  - Ask for hypothesis formation.
- What are the methods used to confirm or reject the hypothesis?
- Whom should you ask to speak?
  - Recognizing slower learners, encouraging all questions.
  - Deciding when to call on quiet group member and ask them a 'softball' question.
  - Moderate the dominant group members.
  - Identify the group members that you can rely on for cogent responses.
  - Supporting and encouraging quiet group members. Tell them how insightful their comments are when they venture to speak!
- Resist the temptation to give a mini lecture.
- Evaluation of learner/discussants. What is the quality of the contributions of individuals to the discussion? What evidence of inquiry and use of educational resources does the contribution of an individual learner/discussant evince? Set your own criteria for the 'ideal' discussion prior to the session (i.e. model discussion outline in ASA PBLD).
- Ability to function on many different levels simultaneously. The single most challenging aspect of moderating/facilitating is that you must monitor the conversation, analyse air time, lead without speaking, evaluate and practice all of the above skills simultaneously while making decisions about them all during the course of the discussion.

8. Conclusion of the session. Watch the clock and leave sufficient time for a strong conclusion to the session.
   - Elements of a strong conclusion.
   - Summary of discussion.
   - Evaluation of issues raised in discussion.
   - Analysis of conclusions and issues that are yet to be resolved.
   - Thanks from the moderator to the discussants.
   - Allow the students to develop the elements of the conclusion. Opportunity for the moderator to dispel misinformation that may have been stated during discussion.

9. Improving your discussion leading skills.
   - Replay the session in your mind and think of other options you might have chosen.
   - Evaluate the quality of the session.
   - Are there any: content omissions, hurt feeling, group ambiance, unresolved issues? Is the preparation of the participants satisfactory or unsatisfactory?
   - Develop strategies to improve future sessions. Expect to have difficulty doing all these moderating/facilitating skills simultaneously on your first attempt!
   - Learn, practice, learn some more, practice, learn again and practice some more.

Appendix 2

Anaesthesia for empyema drainage in a patient with a broncho-pleural fistula*

Objectives

After participating in the discussion the discussant should be able to:

A. Understand techniques employed to avoid pulmonary cross contamination during thoracic surgery.
B. Discuss techniques of anaesthesia induction for lung isolation.
C. Identify endotracheal and endobronchial devices suitable for lung isolation.
D. Manage intra- and post-operative ventilation and analgesia in debilitated thoracic surgery patients.

Scenario

A 60-year-old male with a 40-pack-year smoking history presents for left sided empyema drainage and possible left upper lobe resection for complications stemming from prolonged staphylococcal pneumonia. He has no history, symptoms or signs of cardiac, renal or liver disease. He weighs 130 lbs, has suffered weight loss, is frail, pyrexial (37.8°C), miserable and weak. He coughs up purulent sputum in copious quantities and prefers to sit up in bed. Pulmonary function tests done for insurance purposes prior to his current illness showed mild obstruction but were

*This scenario is reproduced with permission from the American Society of Anesthesiologists and its author, Professor Berend Mets. It is taken from the ASA's PBL refresher session.

unremarkable and showed no improvement with bronchodilators. His blood gases now show decreased oxygenation and normocarbia. He is currently receiving Vancomycin and has no known allergies. His chest X-ray shows a left-sided collection with an air fluid level.

Key questions

1. What is the single most important intervention to minimize the potential for pulmonary cross-contamination?
2. What options for anaesthesia induction for intubation are there? What are the pros and cons of each? What technique would *you* favour?
3. What are the endotracheal and endobronchial devices suitable for lung isolation? Which would *you* use and why?
4. Are there any other issues associated with this patient's disease history which may impact on anesthetic management?
5. How would you adjust ventilation management intra-operatively to combat hypoxia? Would you extubate this patient postoperatively?
6. How would you manage post-operative analgesia?

Model discussion

Pre-operative evaluation

1. Would this patient benefit from further pulmonary function testing. Of what importance is it to personally review the X-rays?

Answer: Location of trachea and bronchus for double lumen tube placement.

Scenario

X-ray shows, pyo-pneumothorax plus parenchymal abcesses and cavitation with air fluid levels. The $PaCO_2$ on room air is 38 mmHg, the PaO_2 is 60 mmHg, the base excess is −3, pH 7.34, bicarbonate 20 MekwIL.

2. What is the relevance of his smoking history? What is the benefit of stopping smoking? What would you advise?

Answer: Decreased CO levels. Less thrombotic risk and cardiac ischaemic risk vs more secretions in first 3 weeks.

Further diagnostic work up

3. Should an underlying neoplasm be sought? Why?

Answer: To help decide whether lung lobe resection is necessary. Look for paraneoplastic syndromes which may result in physiological derangement, hormonal:

- PTH: hypercalcemia
- ADH: water retention
- ACTH: electrolyte disturbance rather than cushingoid picture.

Eaton-Lambert Syndrome: myasthenic weakness associated with succinylcholine.
Hematological abnormalities enhanced thrombosis risk, DVTs etc.

4. What is the post-operative plan, should the patient be ventilated if a lobectomy is performed, why not?

Answer: Avoid positive pressure ventilation effects on bronchial stump: potential for dehiscence especially if septic surgical area.

Pre-operative optimisation

5. Can this patient be optimised? Should surgery be delayed to control temperature or to treat dehydration resulting in mild metabolic acidosis.
   Should the empyema be drained? before induction, if so why, risks? benefits? Should antibiotics be discontinued/changed?

Induction

6. What are the concerns of inducing anesthesia?

Answer: Avoid tension pneumothorax with transfer of pus under pressure to the other lung or passive transfer of pus to R lung when coughing reflex obtunded by anesthesia with or without paralysis as well as when turning patient onto R side for surgery.

With chest tube in place, will paralysis and the need for associated positive pressure ventilation direct all administered gas through the bronchopulmonary fistula (BPF) thus resulting in inadequate gas exchange?

7. Can hemodynamic instability be expected in this patient, due to cachexia with associated hypoalbuminemia, as well as dehydration from elevated temperature and tachypnoea?
8. In the light of this what are the options (advantages/disadvantages) for lung isolation?

Answer: Endobronchial tube or bronchial blockers, to affect lung isolation are an option. However they do not allow non-dependent lung (i.e. diseased lung suctioning and so clearance of secretions, hence there is a chance of contamination post-extubation, and worsening of pneumonia or atelectasis post-operatively as well as enhanced hypoxia intra-operatively.

Single lumen tube down R side, similar problems as also may easily impinge on the right upper lobe and so enhance hypoxia.

Univent tube® allows suctioning (usually inadequate) down 'blocked side' but not ventilation. However when the bronchus is unblocked contamination may then still occur of the other lung, as parynchymal abcess cavities can drain especially to the dependent lung.

Double lumen tube (DLT) placement allows intermittent ventilation and suctioning. It is very important that the left side is clamped before positioning and a working suction catheter placed down the diseased lung side to avoid leakage of pus down tube and so possible contamination via the double lumen tube connector to dependent side upon turning patient on their side for surgery.

Thus protection from lung contamination in the presence of empyema or pus in one lung is dependent on DLT isolation with clamping and diseased lung suctioning to avoid cross contamination especially during patient movement.

Scenario

You have decided to use a Double Lumen Tube.

9. Now what is induction technique? What anesthetic techniques will allow endotracheal placement of the breathing tube?

Answer: A Fibreoptic awake bronchoscopy. *Drawbacks*: inexperience, uncomfortable tachypnoeic patient with copious secretions hence topical local anesthetic not likely to work well. Coughing may cause further rupture of pneumatoceles associated with staphylococcal pneumonia. *Advantages*: cough reflex maintained, spontaneous ventilation maintained till isolation.

Volatile anesthetic induction. *Drawbacks*: difficult to get adequate depth of anesthesia, in light of impaired respiration and gas exchange (may not be true for sevoflurane), obtund cough reflex and so contamination possible. Possible hemodynamic compromise. *Advantages*: maintain spontaneous ventilation till lung isolation. Perhaps combine topical local anesthetic with volatile anesthetic induction.

Rapid sequence intubation, no positive pressure ventilation until placement and isolation of the left lung with lung in dependent position and immediate clamping. Need experience (and luck). *Drawbacks*: inability to place tube immediately would require positive pressure ventilation with the possibility of ventilatory gas loss out of the bronchpleural fistula. If this is small this is not a big issue (usually this isn't a big issue). Possible hemodynamic compromise. *Advantages*: atraumatic to patient early securing of airway, minimal chance of pneumatocele rupture.

Scenario

There is a sudden increase in intrathoracic pressure and pus up tube other side.

10. What can you do?

Maintenance

11. How do you adjust ventilation when one-lung ventilation is started?

Answer: The tidal ventilation can be adjusted in inflation pressures are too high, avoiding baro-trauma and allowing permissive hypercapnea.

12. The patients saturation drops to 90%, but he is hypercarbic. How does this affect your interpretation of this saturation value, what will you do now?

Answer: Show oxygen dissociation curve. In hypercarbia the PaO_2 will be slightly higher than anticipated. Options after assuring DLT appropriately positioned. N.B. Is right upper lobe being ventilated. CPAP, PEEP, or insufflation or tube closure after slight lung inflation.

Scenario

The patient underwent a left lower lobectomy.

13. The surgeon wants you to pull back the left sided double lumen tube, how will you do this?

Answer: Under vision or blindly?

Post-operative care

14. Are you planning to extubate this patient, if so why?

Answer: Extubate at all costs to ensure minimal positive pressure on lobectomy stump in the setting of sepsis.

15. How will you manage post-operative analgesia?

Answer: Epidural, intercostal block, PCA, NSAIDS?

Further reading

Bronchopleural fistula, pulmonary abscess and empyema

Benumof, ed. *Anesthesia for thoracic surgery*, 2nd ed. Anesthesia for Emergency Thoracic Surgery, 1995;626–633.

Barash, Cuiien and Stoelting, eds. *Clinical anesthesia*, 3rd ed. Anesthesia for Thoracic Surgery, 1992;769–803.

Ryder GH, Short DH, Zeitlin GL. The anaesthetic management of patients with bronchopleural fistula with the Robertshaw double-lumen tube. *British Journal of Anaesthesia*. 1965;37:8615.

Riley RH, Wood BM. Induction of anaesthesia in a patient with a bronchopleural fistula (Letter). *Anaesthesia and Intensive Care*. 1994;22:625–626.

Preoperative assessment of pulmonary function for lung resection

Slinger PD, Johnston MR. Preoperative assessment for pulmonary resection. *Journal of Cardiovascular and Thoracic Anaesthesiology*. 2000;14:202–211.

Wyser C, Stultz P, Soler M, et al. Prospective evaluation of an algorithm for the functional assessment of lung resection candidates. *American Journal of Respiratory and Critical Care Medicine*. 2000;159:1450–1456.

Endotracheal and endobronchial intubation

Barash, Cullen and Stoelting, eds. *Clinical anesthesia*, 3rd ed. Anesthesia for Thoracic Surgery, 1992;769–803.

On Interesting Dilemmas in Thoracic Anesthesia

Anesthetic Management of a patient with a descending thoracic aortic aneurysm and severe bilateral bullous pulmonary parenchymal disease. *Canadian Journal of Anaesthesiology*. 1995;427:168–172.
Anaesthesia for patients with mediastinal masses. *Canadian Journal of Anaesthesiology*. 1989;36:681–688.
Cardiac and thoracic vascular injuries. *Southern Medical Journal*. 4:731–739.
Anesthesia for bullectomy. *Anaesthesia*. 1985;40:977–980.
The urinary catheter. *Anaesthesia*. 1983;38:475–477.

Index

Academic competence, 191
Activity of anaesthetic teacher, 94
Adult learner, 9
  Specific needs, 201
Advanced beginner, 211, 214
  Assessment, 214
  Teaching, 211
Advice to trainee, 7
Affective performance, 177
Anaesthetic, 5, 6
  Activity, 94
  Community, 6
  Department, 5, 6
  Education, 197
  Non-technical skills, 54
  Teacher, 8, 94
Analysis, 8
  Performance, 8
Appraisal, 189
Apprenticeship, 27
Assessment, 34, 41, 85, 166
  Clinical performance, 178
  Criterion referenced, 85
  Everyday practice, 175
  Formal, 24, 87
  Format, 170
  In-training, 166
  Ipsative, 85
  Norm-referenced, 4, 85
  Portfolio-based, 173
  Practice observation, 175
  Quality, 35
  Scoring system, 178
  Self, 41
  Simulator, 203
  Summative, 87
Attitude, 23, 59

Backward reasoning, 68
Behaviour, 23, 155
  Difficulties, 162
  How to teach, 160
  Learn, 160
  Professional manner, 155

Clinical, 33, 166
  Competence, 33, 165, 166, 192
  Encounter, 28
  Measure, 40
  Performance, 40
    assessment, 178
  Proficiency, 33
  Skills list, 24, 26
  Supervision, 13
  Teaching, 21, 28
    Curriculum, 21
    Explication, 28
    Reflection, 28
    Review, 28
Communication skills, 7
Competence, 14, 33, 165, 166, 191
  Academic, 191
  Clinical, 33, 166, 192
  Elements, 14
  Judgement, 165, 166
  Professional, 192

Competency, 21, 171
  Controversy, 23
  Criterion-referenced, 171
  Initial test, 14
  Based curriculum, 22
  Based training programme, 14
Competent performer, 22
Components of simulation, 247
Concurrent validity, 36
Consent, 52
Construct validity, 36
Consultant
  Role-model, 106
  Teacher, 106
Context-based scenario, 136
Contextualised knowledge, 85
Criterion referenced assessment, 85
Criterion validity, 36
Criterion-referenced competencies, 171
Curriculum, 21, 27
  Conventional, 63
  Experimental, 27
  PBL, 63
  Simulation, 243

Decision making, 103, 105, 109, 222
  Skills, 103
  Teaching, 109
  Trainee responsibilty, 107
Declaration of Helsinki, 46
Dependency, 10
  Cycle, 10
  Tutor, 10
Duties of the doctor, 157

Educational
  Assessment, 34
  Contract, 75
  Entrepreneur, 74
  Environment, 73
  Measurement, 34
  Project, 5
  Skills, 73
  Supervision, 71
  Supervisor, 71
    Role, 72
Elements of competence, 14
Ethical
  Practice, 50
Ethics of learning, 45
Evaluation, 183
  Process, 183
Examination, 3, 8
  Certification, 8
  Exit, 3
  In-service, 3
  Licensure, 8
Experience
  Cycle, 200
  Plan, 25
Experiential, 200
  Curriculum, 27
  Learning, 200
Explication, 30

Face validity, 36
Fairness, 35
Feedback, 60, 183
  Formal monitoring of progress, 186
  In-theatre, 185
  Monitoring, 183
  Practical, 185
Fire drills, 227, 253
  Obstetric haemorrhage, 253
  Organise, 253
  Preparation, 253
Fixation errors, 221

Good practice, 50
  Principles, 50

Haptic systems, 198
  Virtual reality, 198
Human modelling, 60

Incompetence, 33
Induction day, 144
Information, 111
  Assimilation, 111
  Collection, 111
Informed consent, 49
Initial briefing, 245
Institution
  Academic, 5
  Independent, 5
In-training assessment, 166
Ipsative assessment, 85

Knowledge, 85, 174
 Contextualised, 85
 Factual, 174
 Situated, 85

Learner, 7, 9, 13
 Adult, 9, 10
 Full time, 10
Learning, 1, 6, 7, 45
 Active, 9
 Adult, 201
 Autonomy, 86
 Curve, 124
 Enquiry, 63
 Ethics, 45
 Habits, 10
 Intention, 9
 Levels, 7
 Monitoring, 34
 Motivation, 86
 Needs, 99
 Objectives, 108
 Observation, 22
 Outcomes, 22
 Own, 7
 Pattern, 10
 Planning, 73
 Practice-based, 138
 Principles, 201
 Right environment, 3
 Setting the scene, 1
 Styles, 6
 Target, 7
 Taxonomy, 108
 Teaching, 7
 Types, 174
 Workplace, 96
Lecture format, 63
Logbook, 15, 83

Measure
 Clinical competence, 40
Mentor, 71, 77
 Functions, 79
 Qualities, 78
Mentoring
 Process, 80
 Relationship, 81
Mini-topic, 134
 Content, 137
 Planning, 134
Mini-tutorial, 133
 Informal, 133
 Operating theatre, 133
Mock drill, 254
Monitoring, 187
 Process, 187
 Training, 187

New starter, 144
 Framework, 144
 Practical skills, 146
 Theory, 146
Non-technical skills, 54, 219
 Aptitute, 59
 Key, 103
 Simulation, 219
Norm Referenced Assessment, 85
Novice anaesthetist, 211
 Assessment, 211

Objective, 21
 Learning, 21
 Practical, 21
Observation, 22, 172, 175
 360-degree, 172
 Learning outcome, 22
 Practice, 175
 Work, 176
Operating theatre, 133
 Controversy, 133
 Mini-topics, 133
 Teaching, 93
 Teaching structure, 97

PBL, 63
 Evaluation of student, 64
 Postgraduate education, 66
Performance, 40, 165, 174
 Analysis, 8
 Clinical, 40
 Periodic-reporting, 169
 Scoring, 174
Planning, 134
 Learning, 73
 Mini-topics, 134
 Supervision, 146

Popular teacher, 8
Portfolio, 84
  Benefits, 85
  Formal assessment, 87
  Functions, 84
  Implementation, 88
  Introduction, 88
  Prepare, 85
  Summative assessment, 87
Practical
  Teaching, 7, 121, 127, 130
    Review, 130
  Procedures, 121, 181
  Monitoring progress, 181
  System, 127
  New starter, 146
Practice observation, 170
Practice-based learning, 138
Preceptoral functions, 71
Preliminary reading, 143
Preparing portfolio, 85
Problem based learning, 63
  Cons, 66
  Controversy, 66
  Implementation, 69
  Pros, 66
Procedures, 123
  Competence, 123
Professional, 5, 8, 192
  Competence, 192
  Interactive, 8
  Life, 5
  Relationship, 8
  Task, 33
  Unpleasant experience, 5
Professionalism, 177
Progress, 16
  Dependency, 9
  Independency, 9
  Testing, 16

Qualities, 158
  Mentor, 78
  Practice, 158

Reflection, 29
Reinforcement, 60
Relationship, 155
  Professional, 8
  Workplace, 155
Reliability, 37
Remediation, 191
Review, 28
Role models, 27
Role play, 248
  Examples, 248
Running a session, 246
Running simulation, 245

Scenario, 134
  Context-based, 134
Scoring, 174
  Performance, 174
Script, 244
Simple manikins, 249
Simulated patients, 248
Simulation, 197, 241
  Advantages, 198
  Anaesthesia education, 197
  Components, 247
  Curriculum, 243
  Device, 198
  Educational aims, 242
  Experiential, 200
  High fidelity, 199
  Practicalities, 241
  Types, 198, 247
Simulator, 197, 231
  Advantages, 208
  Centre, 231
  General principles, 202
  Role
    Assessment, 203
    Education, 203
    Research, 203
  Setting up centre, 231
  Scenario, 213
  Technical skills, 208
Situation awareness, 56, 110, 220
  Components, 113
  Steps, 111
Skills, 7, 54, 207
  Acquisition, 207
  Clinical, 25
  Communication, 7, 24
  Decision making, 103
  Developing, 148
  Educational supervisor, 73
  Non-technical, 54
  Technical, 207

Theoratical, 24
Standard, 34, 40
Setting, 34, 40
Supervision, 13
Continuous, 14
Direct, 14
Level, 146, 147
Mechanics, 76
Need, 14
Planning, 146

Task management, 114, 223
Taxonomy, 108
Categories, 108
Learning, 108
Teacher
Anaesthetic, 8
Good, 8
Popular, 8
Teaching, 3, 6, 93
Anaesthesia, 93
Case-based, 7
Clinical, 3
Consent, 52
Decision making skills, 109
Department, 6
Formal, 6
In UK, 6
In USA, 6
New starters, 143
Novice anaesthetist, 143
One to one, 7
Operating theatre, 93
Organissation, 6
Practical procedures, 121
Practical, 7
Scheme, 21
Small group, 6
Topics, 133
Team
Training, 59
Working, 59, 225
Technical skills, 207
Assessment, 207
Simulation, 207
Testing, 16
Progress, 16
Theatre, 94
As classroom, 94
Teaching, 133
Theoretical skills list, 24
Trainee, 5, 8
Difficulty, 5
Learner, 8
Learning needs, 99
Role, 6
Training
Early, 149
Formal, 8
Modular, 128
Objectives, 25
Round the world, 4
Triangulation, 175
Tutor dependency, 10
Types of simulation, 198

Underperformance, 191

Validity, 35
Concurrent, 35
Construct, 35
Content, 35
Criterion, 35
Face, 35
Predictive, 35
Types, 35

Workplace, 96
Learning, 96
Relationship, 155